Time: A Very Short Introduction

VERY SHORT INTRODUCTIONS are for anyone wanting a stimulating and accessible way into a new subject. They are written by experts, and have been translated into more than 45 different languages.

The series began in 1995, and now covers a wide variety of topics in every discipline. The VSI library currently contains over 650 volumes—a Very Short Introduction to everything from Psychology and Philosophy of Science to American History and Relativity—and continues to grow in every subject area.

Very Short Introductions available now:

ABOLITIONISM Richard S. Newman
THE ABRAHAMIC RELIGIONS Charles L. Cohen
ACCOUNTING Christopher Nobes
ADOLESCENCE Peter K. Smith
ADVERTISING Winston Fletcher
AERIAL WARFARE Frank Ledwidge
AESTHETICS Bence Nanay
AFRICAN AMERICAN RELIGION Eddie S. Glaude Jr
AFRICAN HISTORY John Parker and Richard Rathbone
AFRICAN POLITICS Ian Taylor
AFRICAN RELIGIONS Jacob K. Olupona
AGEING Nancy A. Pachana
AGNOSTICISM Robin Le Poidevin
AGRICULTURE Paul Brassley and Richard Soffe
ALEXANDER THE GREAT Hugh Bowden
ALGEBRA Peter M. Higgins
AMERICAN BUSINESS HISTORY Walter A. Friedman
AMERICAN CULTURAL HISTORY Eric Avila
AMERICAN FOREIGN RELATIONS Andrew Preston
AMERICAN HISTORY Paul S. Boyer
AMERICAN IMMIGRATION David A. Gerber
AMERICAN INTELLECTUAL HISTORY Jennifer Ratner-Rosenhagen
AMERICAN LEGAL HISTORY G. Edward White
AMERICAN MILITARY HISTORY Joseph T. Glatthaar
AMERICAN NAVAL HISTORY Craig L. Symonds
AMERICAN POLITICAL HISTORY Donald Critchlow
AMERICAN POLITICAL PARTIES AND ELECTIONS L. Sandy Maisel
AMERICAN POLITICS Richard M. Valelly
THE AMERICAN PRESIDENCY Charles O. Jones
THE AMERICAN REVOLUTION Robert J. Allison
AMERICAN SLAVERY Heather Andrea Williams
THE AMERICAN SOUTH Charles Reagan Wilson
THE AMERICAN WEST Stephen Aron
AMERICAN WOMEN'S HISTORY Susan Ware
AMPHIBIANS T. S. Kemp
ANAESTHESIA Aidan O'Donnell
ANALYTIC PHILOSOPHY Michael Beaney
ANARCHISM Colin Ward
ANCIENT ASSYRIA Karen Radner
ANCIENT EGYPT Ian Shaw
ANCIENT EGYPTIAN ART AND ARCHITECTURE Christina Riggs
ANCIENT GREECE Paul Cartledge

THE ANCIENT NEAR EAST Amanda H. Podany
ANCIENT PHILOSOPHY Julia Annas
ANCIENT WARFARE Harry Sidebottom
ANGELS David Albert Jones
ANGLICANISM Mark Chapman
THE ANGLO-SAXON AGE John Blair
ANIMAL BEHAVIOUR Tristram D. Wyatt
THE ANIMAL KINGDOM Peter Holland
ANIMAL RIGHTS David DeGrazia
THE ANTARCTIC Klaus Dodds
ANTHROPOCENE Erle C. Ellis
ANTISEMITISM Steven Beller
ANXIETY Daniel Freeman and Jason Freeman
THE APOCRYPHAL GOSPELS Paul Foster
APPLIED MATHEMATICS Alain Goriely
THOMAS AQUINAS Fergus Kerr
ARBITRATION Thomas Schultz and Thomas Grant
ARCHAEOLOGY Paul Bahn
ARCHITECTURE Andrew Ballantyne
ARISTOCRACY William Doyle
ARISTOTLE Jonathan Barnes
ART HISTORY Dana Arnold
ART THEORY Cynthia Freeland
ARTIFICIAL INTELLIGENCE Margaret A. Boden
ASIAN AMERICAN HISTORY Madeline Y. Hsu
ASTROBIOLOGY David C. Catling
ASTROPHYSICS James Binney
ATHEISM Julian Baggini
THE ATMOSPHERE Paul I. Palmer
AUGUSTINE Henry Chadwick
AUSTRALIA Kenneth Morgan
AUTISM Uta Frith
AUTOBIOGRAPHY Laura Marcus
THE AVANT GARDE David Cottington
THE AZTECS Davíd Carrasco
BABYLONIA Trevor Bryce
BACTERIA Sebastian G. B. Amyes
BANKING John Goddard and John O. S. Wilson
BARTHES Jonathan Culler
THE BEATS David Sterritt
BEAUTY Roger Scruton
BEHAVIOURAL ECONOMICS Michelle Baddeley
BESTSELLERS John Sutherland
THE BIBLE John Riches
BIBLICAL ARCHAEOLOGY Eric H. Cline
BIG DATA Dawn E. Holmes
BIOCHEMISTRY Mark Lorch
BIOGEOGRAPHY Mark V. Lomolino
BIOGRAPHY Hermione Lee
BIOMETRICS Michael Fairhurst
BLACK HOLES Katherine Blundell
BLASPHEMY Yvonne Sherwood
BLOOD Chris Cooper
THE BLUES Elijah Wald
THE BODY Chris Shilling
THE BOOK OF COMMON PRAYER Brian Cummings
THE BOOK OF MORMON Terryl Givens
BORDERS Alexander C. Diener and Joshua Hagen
THE BRAIN Michael O'Shea
BRANDING Robert Jones
THE BRICS Andrew F. Cooper
THE BRITISH CONSTITUTION Martin Loughlin
THE BRITISH EMPIRE Ashley Jackson
BRITISH POLITICS Tony Wright
BUDDHA Michael Carrithers
BUDDHISM Damien Keown
BUDDHIST ETHICS Damien Keown
BYZANTIUM Peter Sarris
CALVINISM Jon Balserak
ALBERT CAMUS Oliver Gloag
CANADA Donald Wright
CANCER Nicholas James
CAPITALISM James Fulcher
CATHOLICISM Gerald O'Collins
CAUSATION Stephen Mumford and Rani Lill Anjum
THE CELL Terence Allen and Graham Cowling
THE CELTS Barry Cunliffe
CHAOS Leonard Smith
GEOFFREY CHAUCER David Wallace
CHEMISTRY Peter Atkins
CHILD PSYCHOLOGY Usha Goswami

CHILDREN'S LITERATURE Kimberley Reynolds
CHINESE LITERATURE Sabina Knight
CHOICE THEORY Michael Allingham
CHRISTIAN ART Beth Williamson
CHRISTIAN ETHICS D. Stephen Long
CHRISTIANITY Linda Woodhead
CIRCADIAN RHYTHMS Russell Foster and Leon Kreitzman
CITIZENSHIP Richard Bellamy
CITY PLANNING Carl Abbott
CIVIL ENGINEERING David Muir Wood
CLASSICAL LITERATURE William Allan
CLASSICAL MYTHOLOGY Helen Morales
CLASSICS Mary Beard and John Henderson
CLAUSEWITZ Michael Howard
CLIMATE Mark Maslin
CLIMATE CHANGE Mark Maslin
CLINICAL PSYCHOLOGY Susan Llewelyn and Katie Aafjes-van Doorn
COGNITIVE NEUROSCIENCE Richard Passingham
THE COLD WAR Robert J. McMahon
COLONIAL AMERICA Alan Taylor
COLONIAL LATIN AMERICAN LITERATURE Rolena Adorno
COMBINATORICS Robin Wilson
COMEDY Matthew Bevis
COMMUNISM Leslie Holmes
COMPARATIVE LITERATURE Ben Hutchinson
COMPETITION AND ANTITRUST LAW Ariel Ezrachi
COMPLEXITY John H. Holland
THE COMPUTER Darrel Ince
COMPUTER SCIENCE Subrata Dasgupta
CONCENTRATION CAMPS Dan Stone
CONFUCIANISM Daniel K. Gardner
THE CONQUISTADORS Matthew Restall and Felipe Fernández-Armesto
CONSCIENCE Paul Strohm
CONSCIOUSNESS Susan Blackmore
CONTEMPORARY ART Julian Stallabrass
CONTEMPORARY FICTION Robert Eaglestone
CONTINENTAL PHILOSOPHY Simon Critchley
COPERNICUS Owen Gingerich
CORAL REEFS Charles Sheppard
CORPORATE SOCIAL RESPONSIBILITY Jeremy Moon
CORRUPTION Leslie Holmes
COSMOLOGY Peter Coles
COUNTRY MUSIC Richard Carlin
CREATIVITY Vlad Glăveanu
CRIME FICTION Richard Bradford
CRIMINAL JUSTICE Julian V. Roberts
CRIMINOLOGY Tim Newburn
CRITICAL THEORY Stephen Eric Bronner
THE CRUSADES Christopher Tyerman
CRYPTOGRAPHY Fred Piper and Sean Murphy
CRYSTALLOGRAPHY A. M. Glazer
THE CULTURAL REVOLUTION Richard Curt Kraus
DADA AND SURREALISM David Hopkins
DANTE Peter Hainsworth and David Robey
DARWIN Jonathan Howard
THE DEAD SEA SCROLLS Timothy H. Lim
DECADENCE David Weir
DECOLONIZATION Dane Kennedy
DEMENTIA Kathleen Taylor
DEMOCRACY Bernard Crick
DEMOGRAPHY Sarah Harper
DEPRESSION Jan Scott and Mary Jane Tacchi
DERRIDA Simon Glendinning
DESCARTES Tom Sorell
DESERTS Nick Middleton
DESIGN John Heskett
DEVELOPMENT Ian Goldin
DEVELOPMENTAL BIOLOGY Lewis Wolpert
THE DEVIL Darren Oldridge
DIASPORA Kevin Kenny
CHARLES DICKENS Jenny Hartley
DICTIONARIES Lynda Mugglestone
DINOSAURS David Norman

DIPLOMACATIC HISTORY Joseph M. Siracusa
DOCUMENTARY FILM Patricia Aufderheide
DREAMING J. Allan Hobson
DRUGS Les Iversen
DRUIDS Barry Cunliffe
DYNASTY Jeroen Duindam
DYSLEXIA Margaret J. Snowling
EARLY MUSIC Thomas Forrest Kelly
THE EARTH Martin Redfern
EARTH SYSTEM SCIENCE Tim Lenton
ECOLOGY Jaboury Ghazoul
ECONOMICS Partha Dasgupta
EDUCATION Gary Thomas
EGYPTIAN MYTH Geraldine Pinch
EIGHTEENTH-CENTURY BRITAIN Paul Langford
THE ELEMENTS Philip Ball
EMOTION Dylan Evans
EMPIRE Stephen Howe
ENERGY SYSTEMS Nick Jenkins
ENGELS Terrell Carver
ENGINEERING David Blockley
THE ENGLISH LANGUAGE Simon Horobin
ENGLISH LITERATURE Jonathan Bate
THE ENLIGHTENMENT John Robertson
ENTREPRENEURSHIP Paul Westhead and Mike Wright
ENVIRONMENTAL ECONOMICS Stephen Smith
ENVIRONMENTAL ETHICS Robin Attfield
ENVIRONMENTAL LAW Elizabeth Fisher
ENVIRONMENTAL POLITICS Andrew Dobson
ENZYMES Paul Engel
EPICUREANISM Catherine Wilson
EPIDEMIOLOGY Rodolfo Saracci
ETHICS Simon Blackburn
ETHNOMUSICOLOGY Timothy Rice
THE ETRUSCANS Christopher Smith
EUGENICS Philippa Levine
THE EUROPEAN UNION Simon Usherwood and John Pinder
EUROPEAN UNION LAW Anthony Arnull
EVOLUTION Brian and Deborah Charlesworth
EXISTENTIALISM Thomas Flynn
EXPLORATION Stewart A. Weaver
EXTINCTION Paul B. Wignall
THE EYE Michael Land
FAIRY TALE Marina Warner
FAMILY LAW Jonathan Herring
MICHAEL FARADAY Frank A. J. L. James
FASCISM Kevin Passmore
FASHION Rebecca Arnold
FEDERALISM Mark J. Rozell and Clyde Wilcox
FEMINISM Margaret Walters
FILM Michael Wood
FILM MUSIC Kathryn Kalinak
FILM NOIR James Naremore
FIRE Andrew C. Scott
THE FIRST WORLD WAR Michael Howard
FOLK MUSIC Mark Slobin
FOOD John Krebs
FORENSIC PSYCHOLOGY David Canter
FORENSIC SCIENCE Jim Fraser
FORESTS Jaboury Ghazoul
FOSSILS Keith Thomson
FOUCAULT Gary Gutting
THE FOUNDING FATHERS R. B. Bernstein
FRACTALS Kenneth Falconer
FREE SPEECH Nigel Warburton
FREE WILL Thomas Pink
FREEMASONRY Andreas Önnerfors
FRENCH LITERATURE John D. Lyons
FRENCH PHILOSOPHY Stephen Gaukroger and Knox Peden
THE FRENCH REVOLUTION William Doyle
FREUD Anthony Storr
FUNDAMENTALISM Malise Ruthven
FUNGI Nicholas P. Money
THE FUTURE Jennifer M. Gidley
GALAXIES John Gribbin
GALILEO Stillman Drake
GAME THEORY Ken Binmore
GANDHI Bhikhu Parekh
GARDEN HISTORY Gordon Campbell
GENES Jonathan Slack

GENIUS Andrew Robinson
GENOMICS John Archibald
GEOGRAPHY John Matthews and David Herbert
GEOLOGY Jan Zalasiewicz
GEOPHYSICS William Lowrie
GEOPOLITICS Klaus Dodds
GERMAN LITERATURE Nicholas Boyle
GERMAN PHILOSOPHY Andrew Bowie
THE GHETTO Bryan Cheyette
GLACIATION David J. A. Evans
GLOBAL CATASTROPHES Bill McGuire
GLOBAL ECONOMIC HISTORY Robert C. Allen
GLOBAL ISLAM Nile Green
GLOBALIZATION Manfred B. Steger
GOD John Bowker
GOETHE Ritchie Robertson
THE GOTHIC Nick Groom
GOVERNANCE Mark Bevir
GRAVITY Timothy Clifton
THE GREAT DEPRESSION AND THE NEW DEAL Eric Rauchway
HABEAS CORPUS Amanda Tyler
HABERMAS James Gordon Finlayson
THE HABSBURG EMPIRE Martyn Rady
HAPPINESS Daniel M. Haybron
THE HARLEM RENAISSANCE Cheryl A. Wall
THE HEBREW BIBLE AS LITERATURE Tod Linafelt
HEGEL Peter Singer
HEIDEGGER Michael Inwood
THE HELLENISTIC AGE Peter Thonemann
HEREDITY John Waller
HERMENEUTICS Jens Zimmermann
HERODOTUS Jennifer T. Roberts
HIEROGLYPHS Penelope Wilson
HINDUISM Kim Knott
HISTORY John H. Arnold
THE HISTORY OF ASTRONOMY Michael Hoskin
THE HISTORY OF CHEMISTRY William H. Brock
THE HISTORY OF CHILDHOOD James Marten
THE HISTORY OF CINEMA Geoffrey Nowell-Smith
THE HISTORY OF LIFE Michael Benton
THE HISTORY OF MATHEMATICS Jacqueline Stedall
THE HISTORY OF MEDICINE William Bynum
THE HISTORY OF PHYSICS J. L. Heilbron
THE HISTORY OF TIME Leofranc Holford-Strevens
HIV AND AIDS Alan Whiteside
HOBBES Richard Tuck
HOLLYWOOD Peter Decherney
THE HOLY ROMAN EMPIRE Joachim Whaley
HOME Michael Allen Fox
HOMER Barbara Graziosi
HORMONES Martin Luck
HORROR Darryl Jones
HUMAN ANATOMY Leslie Klenerman
HUMAN EVOLUTION Bernard Wood
HUMAN PHYSIOLOGY Jamie A. Davies
HUMAN RIGHTS Andrew Clapham
HUMANISM Stephen Law
HUME James A. Harris
HUMOUR Noël Carroll
THE ICE AGE Jamie Woodward
IDENTITY Florian Coulmas
IDEOLOGY Michael Freeden
THE IMMUNE SYSTEM Paul Klenerman
INDIAN CINEMA Ashish Rajadhyaksha
INDIAN PHILOSOPHY Sue Hamilton
THE INDUSTRIAL REVOLUTION Robert C. Allen
INFECTIOUS DISEASE Marta L. Wayne and Benjamin M. Bolker
INFINITY Ian Stewart
INFORMATION Luciano Floridi
INNOVATION Mark Dodgson and David Gann
INTELLECTUAL PROPERTY Siva Vaidhyanathan
INTELLIGENCE Ian J. Deary
INTERNATIONAL LAW Vaughan Lowe
INTERNATIONAL MIGRATION Khalid Koser

INTERNATIONAL RELATIONS Christian Reus-Smit
INTERNATIONAL SECURITY Christopher S. Browning
IRAN Ali M. Ansari
ISLAM Malise Ruthven
ISLAMIC HISTORY Adam Silverstein
ISLAMIC LAW Mashood A. Baderin
ISOTOPES Rob Ellam
ITALIAN LITERATURE Peter Hainsworth and David Robey
HENRY JAMES Susan L. Mizruchi
JESUS Richard Bauckham
JEWISH HISTORY David N. Myers
JEWISH LITERATURE Ilan Stavans
JOURNALISM Ian Hargreaves
JUDAISM Norman Solomon
JUNG Anthony Stevens
KABBALAH Joseph Dan
KAFKA Ritchie Robertson
KANT Roger Scruton
KEYNES Robert Skidelsky
KIERKEGAARD Patrick Gardiner
KNOWLEDGE Jennifer Nagel
THE KORAN Michael Cook
KOREA Michael J. Seth
LAKES Warwick F. Vincent
LANDSCAPE ARCHITECTURE Ian H. Thompson
LANDSCAPES AND GEOMORPHOLOGY Andrew Goudie and Heather Viles
LANGUAGES Stephen R. Anderson
LATE ANTIQUITY Gillian Clark
LAW Raymond Wacks
THE LAWS OF THERMODYNAMICS Peter Atkins
LEADERSHIP Keith Grint
LEARNING Mark Haselgrove
LEIBNIZ Maria Rosa Antognazza
C. S. LEWIS James Como
LIBERALISM Michael Freeden
LIGHT Ian Walmsley
LINCOLN Allen C. Guelzo
LINGUISTICS Peter Matthews
LITERARY THEORY Jonathan Culler
LOCKE John Dunn
LOGIC Graham Priest
LOVE Ronald de Sousa
MARTIN LUTHER Scott H. Hendrix
MACHIAVELLI Quentin Skinner
MADNESS Andrew Scull
MAGIC Owen Davies
MAGNA CARTA Nicholas Vincent
MAGNETISM Stephen Blundell
MALTHUS Donald Winch
MAMMALS T. S. Kemp
MANAGEMENT John Hendry
NELSON MANDELA Elleke Boehmer
MAO Delia Davin
MARINE BIOLOGY Philip V. Mladenov
MARKETING Kenneth Le Meunier-FitzHugh
THE MARQUIS DE SADE John Phillips
MARTYRDOM Jolyon Mitchell
MARX Peter Singer
MATERIALS Christopher Hall
MATHEMATICAL FINANCE Mark H. A. Davis
MATHEMATICS Timothy Gowers
MATTER Geoff Cottrell
THE MAYA Matthew Restall and Amara Solari
THE MEANING OF LIFE Terry Eagleton
MEASUREMENT David Hand
MEDICAL ETHICS Michael Dunn and Tony Hope
MEDICAL LAW Charles Foster
MEDIEVAL BRITAIN John Gillingham and Ralph A. Griffiths
MEDIEVAL LITERATURE Elaine Treharne
MEDIEVAL PHILOSOPHY John Marenbon
MEMORY Jonathan K. Foster
METAPHYSICS Stephen Mumford
METHODISM William J. Abraham
THE MEXICAN REVOLUTION Alan Knight
MICROBIOLOGY Nicholas P. Money
MICROECONOMICS Avinash Dixit
MICROSCOPY Terence Allen
THE MIDDLE AGES Miri Rubin
MILITARY JUSTICE Eugene R. Fidell
MILITARY STRATEGY Antulio J. Echevarria II
MINERALS David Vaughan
MIRACLES Yujin Nagasawa
MODERN ARCHITECTURE Adam Sharr

MODERN ART David Cottington
MODERN BRAZIL Anthony W. Pereira
MODERN CHINA Rana Mitter
MODERN DRAMA
Kirsten E. Shepherd-Barr
MODERN FRANCE
Vanessa R. Schwartz
MODERN INDIA Craig Jeffrey
MODERN IRELAND Senia Pašeta
MODERN ITALY Anna Cento Bull
MODERN JAPAN
Christopher Goto-Jones
MODERN LATIN AMERICAN LITERATURE
Roberto González Echevarría
MODERN WAR Richard English
MODERNISM Christopher Butler
MOLECULAR BIOLOGY
Aysha Divan and Janice A. Royds
MOLECULES Philip Ball
MONASTICISM Stephen J. Davis
THE MONGOLS Morris Rossabi
MONTAIGNE William M. Hamlin
MOONS David A. Rothery
MORMONISM
Richard Lyman Bushman
MOUNTAINS Martin F. Price
MUHAMMAD Jonathan A. C. Brown
MULTICULTURALISM Ali Rattansi
MULTILINGUALISM John C. Maher
MUSIC Nicholas Cook
MYTH Robert A. Segal
NAPOLEON David Bell
THE NAPOLEONIC WARS
Mike Rapport
NATIONALISM Steven Grosby
NATIVE AMERICAN LITERATURE
Sean Teuton
NAVIGATION Jim Bennett
NAZI GERMANY Jane Caplan
NEOLIBERALISM Manfred B. Steger and Ravi K. Roy
NETWORKS Guido Caldarelli and Michele Catanzaro
THE NEW TESTAMENT
Luke Timothy Johnson
THE NEW TESTAMENT AS LITERATURE Kyle Keefer
NEWTON Robert Iliffe
NIELS BOHR J. L. Heilbron
NIETZSCHE Michael Tanner
NINETEENTH-CENTURY BRITAIN
Christopher Harvie and H. C. G. Matthew
THE NORMAN CONQUEST
George Garnett
NORTH AMERICAN INDIANS
Theda Perdue and Michael D. Green
NORTHERN IRELAND
Marc Mulholland
NOTHING Frank Close
NUCLEAR PHYSICS Frank Close
NUCLEAR POWER Maxwell Irvine
NUCLEAR WEAPONS
Joseph M. Siracusa
NUMBER THEORY Robin Wilson
NUMBERS Peter M. Higgins
NUTRITION David A. Bender
OBJECTIVITY Stephen Gaukroger
OCEANS Dorrik Stow
THE OLD TESTAMENT
Michael D. Coogan
THE ORCHESTRA D. Kern Holoman
ORGANIC CHEMISTRY
Graham Patrick
ORGANIZATIONS Mary Jo Hatch
ORGANIZED CRIME
Georgios A. Antonopoulos and Georgios Papanicolaou
ORTHODOX CHRISTIANITY
A. Edward Siecienski
OVID Llewelyn Morgan
PAGANISM Owen Davies
PAIN Rob Boddice
THE PALESTINIAN-ISRAELI CONFLICT Martin Bunton
PANDEMICS Christian W. McMillen
PARTICLE PHYSICS Frank Close
PAUL E. P. Sanders
PEACE Oliver P. Richmond
PENTECOSTALISM William K. Kay
PERCEPTION Brian Rogers
THE PERIODIC TABLE Eric R. Scerri
PHILOSOPHICAL METHOD
Timothy Williamson
PHILOSOPHY Edward Craig
PHILOSOPHY IN THE ISLAMIC WORLD Peter Adamson
PHILOSOPHY OF BIOLOGY
Samir Okasha

PHILOSOPHY OF LAW
Raymond Wacks
PHILOSOPHY OF PHYSICS
David Wallace
PHILOSOPHY OF SCIENCE
Samir Okasha
PHILOSOPHY OF RELIGION
Tim Bayne
PHOTOGRAPHY Steve Edwards
PHYSICAL CHEMISTRY Peter Atkins
PHYSICS Sidney Perkowitz
PILGRIMAGE Ian Reader
PLAGUE Paul Slack
PLANETS David A. Rothery
PLANTS Timothy Walker
PLATE TECTONICS Peter Molnar
PLATO Julia Annas
POETRY Bernard O'Donoghue
POLITICAL PHILOSOPHY David Miller
POLITICS Kenneth Minogue
POPULISM Cas Mudde and Cristóbal Rovira Kaltwasser
POSTCOLONIALISM Robert Young
POSTMODERNISM Christopher Butler
POSTSTRUCTURALISM
Catherine Belsey
POVERTY Philip N. Jefferson
PREHISTORY Chris Gosden
PRESOCRATIC PHILOSOPHY
Catherine Osborne
PRIVACY Raymond Wacks
PROBABILITY John Haigh
PROGRESSIVISM Walter Nugent
PROHIBITION W. J. Rorabaugh
PROJECTS Andrew Davies
PROTESTANTISM Mark A. Noll
PSYCHIATRY Tom Burns
PSYCHOANALYSIS Daniel Pick
PSYCHOLOGY Gillian Butler and Freda McManus
PSYCHOLOGY OF MUSIC
Elizabeth Hellmuth Margulis
PSYCHOPATHY Essi Viding
PSYCHOTHERAPY Tom Burns and Eva Burns-Lundgren
PUBLIC ADMINISTRATION
Stella Z. Theodoulou and Ravi K. Roy
PUBLIC HEALTH Virginia Berridge
PURITANISM Francis J. Bremer
THE QUAKERS Pink Dandelion
QUANTUM THEORY
John Polkinghorne
RACISM Ali Rattansi
RADIOACTIVITY Claudio Tuniz
RASTAFARI Ennis B. Edmonds
READING Belinda Jack
THE REAGAN REVOLUTION Gil Troy
REALITY Jan Westerhoff
RECONSTRUCTION Allen C. Guelzo
THE REFORMATION Peter Marshall
REFUGEES Gil Loescher
RELATIVITY Russell Stannard
RELIGION Thomas A. Tweed
RELIGION IN AMERICA Timothy Beal
THE RENAISSANCE Jerry Brotton
RENAISSANCE ART
Geraldine A. Johnson
RENEWABLE ENERGY Nick Jelley
REPTILES T. S. Kemp
REVOLUTIONS Jack A. Goldstone
RHETORIC Richard Toye
RISK Baruch Fischhoff and John Kadvany
RITUAL Barry Stephenson
RIVERS Nick Middleton
ROBOTICS Alan Winfield
ROCKS Jan Zalasiewicz
ROMAN BRITAIN Peter Salway
THE ROMAN EMPIRE Christopher Kelly
THE ROMAN REPUBLIC
David M. Gwynn
ROMANTICISM Michael Ferber
ROUSSEAU Robert Wokler
RUSSELL A. C. Grayling
THE RUSSIAN ECONOMY
Richard Connolly
RUSSIAN HISTORY Geoffrey Hosking
RUSSIAN LITERATURE Catriona Kelly
THE RUSSIAN REVOLUTION
S. A. Smith
SAINTS Simon Yarrow
SAMURAI Michael Wert
SAVANNAS Peter A. Furley
SCEPTICISM Duncan Pritchard
SCHIZOPHRENIA Chris Frith and Eve Johnstone
SCHOPENHAUER
Christopher Janaway
SCIENCE AND RELIGION
Thomas Dixon
SCIENCE FICTION David Seed

THE SCIENTIFIC REVOLUTION Lawrence M. Principe
SCOTLAND Rab Houston
SECULARISM Andrew Copson
SEXUAL SELECTION Marlene Zuk and Leigh W. Simmons
SEXUALITY Véronique Mottier
WILLIAM SHAKESPEARE Stanley Wells
SHAKESPEARE'S COMEDIES Bart van Es
SHAKESPEARE'S SONNETS AND POEMS Jonathan F. S. Post
SHAKESPEARE'S TRAGEDIES Stanley Wells
GEORGE BERNARD SHAW Christopher Wixson
SIKHISM Eleanor Nesbitt
SILENT FILM Donna Kornhaber
THE SILK ROAD James A. Millward
SLANG Jonathon Green
SLEEP Steven W. Lockley and Russell G. Foster
SMELL Matthew Cobb
ADAM SMITH Christopher J. Berry
SOCIAL AND CULTURAL ANTHROPOLOGY John Monaghan and Peter Just
SOCIAL PSYCHOLOGY Richard J. Crisp
SOCIAL WORK Sally Holland and Jonathan Scourfield
SOCIALISM Michael Newman
SOCIOLINGUISTICS John Edwards
SOCIOLOGY Steve Bruce
SOCRATES C. C. W. Taylor
SOFT MATTER Tom McLeish
SOUND Mike Goldsmith
SOUTHEAST ASIA James R. Rush
THE SOVIET UNION Stephen Lovell
THE SPANISH CIVIL WAR Helen Graham
SPANISH LITERATURE Jo Labanyi
SPINOZA Roger Scruton
SPIRITUALITY Philip Sheldrake
SPORT Mike Cronin
STARS Andrew King
STATISTICS David J. Hand
STEM CELLS Jonathan Slack
STOICISM Brad Inwood
STRUCTURAL ENGINEERING David Blockley
STUART BRITAIN John Morrill
THE SUN Philip Judge
SUPERCONDUCTIVITY Stephen Blundell
SUPERSTITION Stuart Vyse
SYMMETRY Ian Stewart
SYNAESTHESIA Julia Simner
SYNTHETIC BIOLOGY Jamie A. Davies
SYSTEMS BIOLOGY Eberhard O. Voit
TAXATION Stephen Smith
TEETH Peter S. Ungar
TELESCOPES Geoff Cottrell
TERRORISM Charles Townshend
THEATRE Marvin Carlson
THEOLOGY David F. Ford
THINKING AND REASONING Jonathan St B. T. Evans
THOUGHT Tim Bayne
TIBETAN BUDDHISM Matthew T. Kapstein
TIDES David George Bowers and Emyr Martyn Roberts
TIME Jenann Ismael
TOCQUEVILLE Harvey C. Mansfield
LEO TOLSTOY Liza Knapp
TOPOLOGY Richard Earl
TRAGEDY Adrian Poole
TRANSLATION Matthew Reynolds
THE TREATY OF VERSAILLES Michael S. Neiberg
TRIGONOMETRY Glen Van Brummelen
THE TROJAN WAR Eric H. Cline
TRUST Katherine Hawley
THE TUDORS John Guy
TWENTIETH-CENTURY BRITAIN Kenneth O. Morgan
TYPOGRAPHY Paul Luna
THE UNITED NATIONS Jussi M. Hanhimäki
UNIVERSITIES AND COLLEGES David Palfreyman and Paul Temple
THE U.S. CIVIL WAR Louis P. Masur
THE U.S. CONGRESS Donald A. Ritchie
THE U.S. CONSTITUTION David J. Bodenhamer
THE U.S. SUPREME COURT Linda Greenhouse
UTILITARIANISM Katarzyna de Lazari-Radek and Peter Singer

UTOPIANISM Lyman Tower Sargent
VETERINARY SCIENCE James Yeates
THE VIKINGS Julian D. Richards
THE VIRTUES Craig A. Boyd and Kevin Timpe
VIRUSES Dorothy H. Crawford
VOLCANOES Michael J. Branney and Jan Zalasiewicz
VOLTAIRE Nicholas Cronk
WAR AND RELIGION Jolyon Mitchell and Joshua Rey
WAR AND TECHNOLOGY Alex Roland
WATER John Finney
WAVES Mike Goldsmith
WEATHER Storm Dunlop
THE WELFARE STATE David Garland
WITCHCRAFT Malcolm Gaskill
WITTGENSTEIN A. C. Grayling
WORK Stephen Fineman
WORLD MUSIC Philip Bohlman
THE WORLD TRADE ORGANIZATION Amrita Narlikar
WORLD WAR II Gerhard L. Weinberg
WRITING AND SCRIPT Andrew Robinson
ZIONISM Michael Stanislawski
ÉMILE ZOLA Brian Nelson

Available soon:

PHILOSOPHY OF MIND Barbara Gail Montero
INSECTS Simon Leather
PAKISTAN Pippa Virdee
THE ARCTIC Klaus Dodds and Jamie Woodward
JAMES JOYCE Colin MacCabe

For more information visit our website

www.oup.com/vsi/

Jenann Ismael

TIME

A Very Short Introduction

OXFORD
UNIVERSITY PRESS

Great Clarendon Street, Oxford, OX2 6DP,
United Kingdom

Oxford University Press is a department of the University of Oxford.
It furthers the University's objective of excellence in research, scholarship,
and education by publishing worldwide. Oxford is a registered trade mark of
Oxford University Press in the UK and in certain other countries

First edition published in 2021

Published in the United States of America by Oxford University Press
198 Madison Avenue, New York, NY 10016, United States of America

British Library Cataloguing in Publication Data
Data available

Library of Congress Control Number: 2021936812

ISBN 978–0–19–883266–9

Printed and bound by CPI Group (UK) Ltd, Croydon, CR0 4YY

Preface

A moment of a human life, a thin slice carved surgically out of the ongoing flow of your being, will have a temporal content that belies its lack of breadth. Each moment of your life is a sort of glimpse of the whole, a tiny capsule that contains both a backward-looking image of its past and a forward-looking image for its future, singled out by its place at the cusp of a particular transition from future into past.

And yet. Time in the world, time out in the wild, time before there were people to capture and preserve it, to prefigure and post-judge it, to layer each moment with memories of its past and hopes for its future. Time itself is just one dimension of a four-dimensional manifold of events. There are differences between the spatial and temporal dimensions but they are subtle, buried in what physicists call the signature of the metric. Time as it appears in the physicist's image of the world does not pass or unfold and does not come with any built-in asymmetry. There is no more difference between past and future than there is between east and west. And the moments of time are themselves indistinguishable and interchangeable.

This is a book about what physics teaches us about time, the line of thought that starts with Newton and leads through the Theory of Relativity. It is also inseparably about bridging the chasm between the physicist's understanding of time, and that of human beings like you and I, embedded in time and living human lives.

The history of physics from Newton, through his debate with Leibniz, to Einstein's two revolutions wrought changes in our conception of time that we couldn't have anticipated from the armchair. One hears nowadays that physics is in a state of philosophical confusion. That is true of some parts of physics, but this strand of development—the one leading from Newton to Einstein and culminating in the general theory of relativity—is a story of great conceptual beauty and philosophical illumination, illumination that could never have been achieved by purely philosophical methods.

The discussion here will provide an opportunity to reflect on what distinguishes the methods of physics from those of philosophy. You will see those methods in evidence throughout this book, displayed with the kind of hindsight that discards all of the false starts and dead ends, replacing the wayward path with a straight shot to the destination.

Since time comes into physics by way of its connection to motion, we will spend a good deal of time talking about the physical theories that were developed to describe how things move. Chapters 1 and 2 introduce the theoretical developments in the 17th to the 20th centuries that brought us from Newtonian mechanics to general relativity. In Chapter 3, we will begin exploring the philosophical implications of the resulting image of time. We will look at some of the strange consequences of relativity, from the way time seems to stretch or shrink depending on one's state of motion, to questions about the possibility of time travel. In Chapter 4 we will raise a puzzle that will occupy the remainder of the book. It is a puzzle about the relationship between time as it appears in physics—a mathematical abstraction with no direction or movement—and the familiar flowing time of human life, which is as messy, tension laden, and fraught as human life itself: full of suspense and action, surprise and regret.

I've put emphasis on the most interesting and philosophically intriguing parts of all of this, the parts that are still waiting to be fully assimilated and fully resolved, and the parts that make the most direct contact with time as we know it. Those are also the parts that have the most to teach us about ourselves and our place in the cosmos.

This introduction is for someone who wants to dip her toe into the challenging and alien waters of what physics has taught us about time, to raise questions such as: What is the difference between time and space? What is the difference between the past and future? Does time flow? Is time travel possible? Is the passage of time an illusion? It is the kind of book that you can read with no prior knowledge and come away with a qualitative sense of what physics seems to be suggesting, not just about the *t* parameter in mathematical formulae tracking motion, but time as you encounter it in your experience: the time of poetry and literature, the time of life itself. It is unabashedly and without embarrassment for the *non*-insider, so I have presupposed nothing, omitted technical discussion that wouldn't be helpful, avoided topics that only professional philosophers or physicists would be interested in, and haven't been shy about using helpful analogies and metaphors. It is possible to get a qualitative understanding of the physical theories discussed here, and the reasoning that led to them, without going all the way into the weeds with the number crunching, experiments, and empirical reasoning it took to produce them. That doesn't mean that it is easy, however. The philosophical ideas involved are difficult and abstract and quite beautiful. I hope they will leave you with an appetite for more and if they do, there is a whole treasure chest of books that immerse you in different aspects of the subject that I can heartily recommend, and which you will find listed in the Further Reading section.

Contents

List of illustrations

Chapter 1
Time until Newton

The invention of time

Until the late 4th century BCE, when people wanted to talk about a past event they had various more or less roundabout ways of locating it in time. They could relate it to some landmark occurrence like a famous battle or solar eclipse. It could be dated by giving the name of the holder of an annual office of state. You could say, for example, that something happened when so and so was chief magistrate. In kingdoms, it was common to use the coronation year of the monarch. A child could be born in the fifth year of Alexander the Great or the fourth year of Nabonidas. One could combine the two, saying for example that something happened in the spring of the year when King Hammurabi destroyed the city of Mari, or in the third year of King Enlil-bani before the great dust storm.

The idea is to use named eras and publicly known events as points of reference to narrow down the location of an event in time. It's the same kind of thing you do at a family dinner when you try to work out when Shereen went through her wearing-ski-gloves-to-bed phase (it was while we were still living at the lake, before we had the blue Volvo, right? The summer we all played ZimZam). There are obvious limitations to the method. It relies on common knowledge of the landmark events you are using as points of reference. No one outside your family would be helped by knowing something happened the summer you were playing ZimZam, for example. Since common knowledge of public events varied from one area to another, so too would the dating and placing systems in use. This method was parochial by its nature.

So long as most communication was local, this wasn't too much of a problem, but as soon as people started having commerce with folks who live at a geographical distance, its limitations became glaring. It is non-trivial to renegotiate a common system of temporal reference whenever meeting people who come from a different city or state. It was, moreover, both cumbersome and imprecise. Consider how the Greek historian Thucydides dated the outbreak of the Peloponnesian War:

> The Thirty Years Treaty agreed after the conquest of Euboea lasted for fourteen years. In the fifteenth year, when Chrysis was in her forty-eighth year as priestess at Argos, Aenesias was ephor in Sparta, Pythodorus had two more months of his archonship in Athens, in the sixth month after the battle of Potidaea, and at the beginning of spring, in the first watch of the night an armed force of slightly over three hundred Thebans entered Plataea, a city in Boeotia allied to Athens.

Eventually someone had a better idea. It happened in the political turmoil that followed the death of Alexander the Great in Babylon in 323 BCE. One of Alexander's Macedonian generals introduced a new system for keeping track of time that became the ancestor of time-keeping systems of every subsequent era. It began from Year 1 at Seleucus I Nicator's arrival in Babylon (what we would call spring 311 BCE) and continued without interruption after his death. His son and successors let the clock run after his death and it became the world's first universal, continuous, and irreversible tally of the passing years. Time was now marked by a number that never restarted, wasn't tied to political events, to the lifecycle of rulers, or limited to geographical regions.

Where before we used to use one event to locate another, the new dating system gave us a kind of transcendent grid extending indefinitely to the past and future, on which every event has a place. It didn't just provide a uniform way of referring to events in the past, it also made the future seem more concrete and definite, like a place where things happened. Numbers function as names for locations on the grid not only because they are easy to remember, but because they come with an order that we can exploit. Assigning numbers to events in a way that reflects the order of the events to which they are assigned allows us to read the order of the events off the order of their dates. Now, instead of just creating timelines for local events, people could coordinate the more parochial placing and dating systems with one another by mapping them into the grid. Time itself now had a structure and events had a place in time.

Physics

Let's jump ahead to the 17th century. Although there had long been a philosophical discussion concerning the nature of time, something quite special happened when the question passed into the hands of the physicists. Up until the 17th century, the people who asked the questions that we think of as distinctive of modern physics were philosophers. They were people like Thales and Lucretius, Anaximander and Aristotle. They built philosophical systems that told us what different kinds of things there were in the world, how those could be assembled, and the principles that governed their movements and behaviours. Thales held that all matter is water, Lucretius that everything is atoms in the void, and Anaximander that the Boundless is the origin of all that is.

Historians disagree about when the so-called scientific revolution began and how continuous it was with the preceding history, but they agree that part of what made it possible for science to emerge as a distinctive enterprise was social developments in Europe that gave rise to scientific societies.

These societies regulated the accumulation and exchange of observational information and that meant that for the first time in history gathering evidence became a collective and systematic enterprise. The commonly observed qualitative regularities that formed the basis of ancient worldviews were replaced by a large fund of carefully gathered information. People were no longer content to simply observe but began to measure. Experiments were performed, results were shared, and theories were published. With the availability of large bodies of data, mathematics acquired a new and prominent role. Now instead of basing theories on the manifest regularities of the everyday world, scientists began to notice more abstract regularities hidden in the data. Tools were developed to search for these hidden regularities and physics as we know it was born.

The Aristotelian worldview, which was the dominant view of the world from the 3rd century BCE to the 17th century, was a comfortable and homey place for a human being. The universe was finite and the Earth was at its centre. A sphere made of a crystalline substance that contained the fixed stars revolved around the Earth. Below the star-encrusted crystal of the sky there were four elements—water, earth, fire, and air. Each of these had a natural motion that described its movements when unimpeded. Water and earth moved towards the centre of the Earth; fire and air, away from it. This view of the world formed the backdrop for education and scholarship in the western world for two millennia, and it was the one into which Isaac Newton, an 18-year-old student who entered Cambridge in 1661, was educated. The plague affecting Cambridge sent young Newton back to his family home in Woolsthorpe twice in his college years, and it was during these visits that he had the insights that led to the majestic theory of motion. The theory was published in 1687 in three volumes presented to the Royal Society of London under the title *Philosophiæ Naturalis Principia Mathematica* (Mathematical Principles of Natural Philosophy).

This was the first physical theory in the modern sense and it undoes the Aristotelian universe in every aspect. Where Aristotle's universe was finite and had a centre, Newton's universe is infinite and has no centre. Where Aristotle's universe divided the heavens from the Earth, in Newton's theory celestial bodies are made of the same stuff and obey the same laws as those on Earth. According to the Newtonian conception of the world, the universe consists entirely of material particles. Objects, from tables to trees and penguins to planets, are configurations of such particles. The behaviour of those configurations is determined by the behaviours of the particles of which they are made. The laws that govern those behaviours are mathematical principles that are the same always and everywhere.

You may have heard of the proverbial 'aha' moment Newton had in his garden seeing an apple fall from a tree and surmising that the force drawing the apple to the ground is the same one keeping the planets in orbit. That insight had the profoundest effects on what it is possible to know about the universe. It made it possible to study the laws that govern planetary motion by studying the motions of things close to the surface of the Earth.

According to Newton's theory, the only feature of material particles that can change with time is position. There are two laws that govern changes of position:

1. a body with no forces applied to it remains at rest if it is at rest and continues moving uniformly in a straight line if it is moving;
2. the force applied to an object is equal to its mass multiplied by its acceleration.

These laws are so strong that if we are given a list of the positions of all of the particles in the world at any particular time, together with their velocities (i.e. the speed and direction in which position is changing), we can calculate their positions at all other times.

This is sometimes expressed by saying that to create the world, all God had to do was put down the particles of which the universe is made, specify their velocities, and decree these two laws. Once those things were done, everything else would follow (as they say) like clockwork.

The idea that it would be possible to write down mathematical principles that are so powerful that they would determine the complete history of the universe down to its finest detail—everything from the way that a particular drop of water lands on the petal of a Siberian peony in the 3rd century BCE to the path the butterfly will take as it flutters across your garden long after you are dust and ashes—from a specification of the initial positions and velocities of the particles that make it up is astounding. And to think that it could be set out in two lines as simple and straightforward as those above seems outlandish. Newton didn't just write down these laws. In the three volumes of his *Principia*, he showed how to use them to derive all known motions of objects, from rocks rolling down mountainsides and balls swinging from pendula, to the motions of planets. The scope and beauty of what he achieved is hard to overemphasize.

Philosophical dispute

Understanding what physics tells us about time is inseparable from understanding what it tells us about space. Motion is change of place in time, so both are part and parcel of theories of motion. Newton had some strong views about space and time that were expressed in an essay that was attached to the *Principia* as a philosophical addendum. In that addendum, he affirmed the view that space and time were *things* that were distinct from the material objects and events that were located in them. In saying this, he was opposing the position of the German philosopher and polymath Gottfried Wilhelm Leibniz. Leibniz was well known to Newton; he had discovered calculus independently around the same time as Newton and they were rivals in the intellectual world of the time. Leibniz denied that space and time were things in their own right. He held instead that what we call space and time were nothing but abstract frameworks for representing relations among objects and events. The disagreement between the two views led to a lively back and forth in an exchange of letters between Leibniz and Samuel Clarke, an English supporter of Newton's. The letters were exchanged in 1715 and 1716 and published the next year. They remain a *locus classicus* of philosophical discussion about the nature of space and time.

Take a moment to look around and take stock of your surroundings and ask yourself whether you see the locations of things in space. You might be inclined to say that you do, but on reflection you will realize that that is not quite right. What you see is actually their spatial relations to the other objects in view. If someone took all of the objects in space—you included—and, without disturbing their relations to one another, moved the whole system some fixed distance or rotated everything around a fixed axis and then laid them down again, you wouldn't notice any difference. Your own body would have to be included in the system so that your own relationships to objects were maintained under the motion, but so long as they remained undisturbed, you wouldn't notice a thing.

Ask yourself now how you know the locations of events in time. We have all kinds of ways of getting our temporal bearings. You can look at your watch, the position of the Sun, or what's on the TV. In all of these cases, however, what we are really perceiving isn't the locations of events in time, but the temporal relations among them. You know that something happened at noon, because your watch registers 12 the moment it occurred. If someone took all the events in history and, keeping their relations to one another intact, moved them backward or forward in time by some fixed interval, you wouldn't be able to tell the difference. The Sun would still rise as the rooster crows. The news would come on as the Sun begins to set, the tide would still rise and fall with the phases of the Moon, and the rhythms of life would beat in the same tempo as the hands wound round the clock. We could even alter the *amount* of time that passes between two events. Since we tell how much time has passed by counting days or watching the hands go around the clock, we could stretch and shrink the interval ticked out by our clocks and so long as we kept the relative durations of physical processes fixed, we would not be able to tell the difference.

Newton thought that space was a thing in its own right and that even if we couldn't see it directly, there was a fact about how an object or system of objects was embedded in space. And likewise, he thought time was a thing in its own right and that even if we couldn't see it directly, there was a fact about how an event (or system of events) was entered in time. That meant that he was committed to the idea that there are infinitely many different ways the world could be that we can't distinguish. There was one for each of the infinite number of ways that objects could be moved or rotated in space without disturbing their relations to one another. And there was one for each of the infinite number of ways we could move events forward or backward in time, or stretch the intervals between them, without disturbing their relations to one another.

Leibniz thought that these considerations provided strong arguments in favour of his own view that space and time were not things in themselves. If space and time were abstract frameworks for representing the spatial and temporal relationships among things, once you specified the spatial relations among events, there was no further fact about how they were entered into space. And once you specified the temporal relations among events, there was no further fact about how they were entered into time. Leibniz's case seemed to receive further support from a curious fact that was first noticed by Galileo, the 16th-century Italian astronomer who invented the first telescope, discovered the law of the pendulum, and was famously convicted for suspicion of heresy for his defence of the Copernican sun-centred astronomy. Galileo argued that if we took a whole system of material bodies and, maintaining their positions relative to one another, didn't just move them and put them down, but rather *set them in motion* and watched how things behaved, then as long as they were moving at uniform velocity, we wouldn't notice a thing. Everything would behave just as it does at rest. Galileo proposes an experiment:

> Shut yourself up with some friend in the main cabin below decks on some large ship, and have with you there some flies, butterflies and other small flying animals. Have a large bowl of water with some fish in it: hang up a bottle that empties drop by drop into a wide vessel beneath it. With the ship standing still, observe carefully how the little animals fly with equal speed to all sides of the cabin. The fish swim differently in all directions; the drop falls into the vessel beneath; and, in throwing something to your friend, you need throw no more strongly in one direction than another, the distances being equal; jumping with your feet together, you pass equal spaces in every direction. When you have observed all these things carefully (though there is no doubt that when the ship is standing still everything must happen in this way), have the ship proceed with any speed you like, so long as the motion is uniform and not fluctuating this way and that. You will discover not the least change in all the effects named, nor could you tell from any of them whether the ship was moving or standing still.

He is correct. So long as the ship is moving at a constant velocity all experiments carried out in the two states of the ship would have the same observable outcome and so there would be no way for a person whose experience was confined to what was going on below decks on the ship to know whether it was moving. This goes not only for the actions he describes—water dropping, balls being thrown, people jumping up and down—but any experiment whatsoever. It turns out to be an explicitly derivable property of Newton's laws that there is no physical process that would happen any differently on Galileo's ship. And it follows from this that according to Newton's own theory, motion at a constant velocity is just as undetectable as position in space or time.

This is exactly what one would expect if Leibniz was right, and so you might think he has the upper hand in his debate with Newton. Why should anyone think that space and time were things, rather than just a network of spatial and temporal relationships among things? Space and time, considered as things in their own right, are not observable. Even motion at a constant velocity (which is change in position over time) makes no difference to the movements or behaviours of anything that we can see. If this were the end of the story, then Leibniz's position might have won the day. But things are not so simple. It turns out that while we can't tell whether an object is moving at a constant velocity, we *can* tell whether it is accelerating. (Although ordinary language uses 'acceleration' to mean speeding up, in physics it means any change in velocity, so slowing down is a form of acceleration since it involves a change in speed, and rotation is a form of acceleration because a rotating object is always changing direction; a rotating object speeds up, effectively, in the direction of the change.) Newton's laws predict, and experience confirms, that things behave differently when they are accelerating from when they are moving at a constant velocity. There are innumerable examples. You don't leave a drink on your tray table while a plane is taking off, for example, because it is apt to slide towards your lap in the acceleration of take-off, although it is fine once the plane reaches cruising speed. Newton himself gave two simple and vivid illustrations. If you take a bucket filled with water and hang it from the ceiling, the surface of the water will lie flat. But if you twist the rope from which it hangs very tight and then let it go, the bucket will rotate, and you will see the water recede from the middle of the bucket and rise up the sides so that the surface makes a U-shape (Figure 1).

If we just confine our attention to the bucket, the relative position of the bucket and the water it contains doesn't change. The only

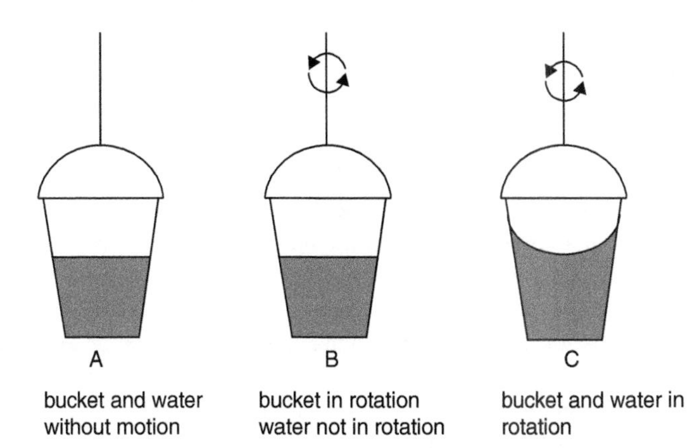

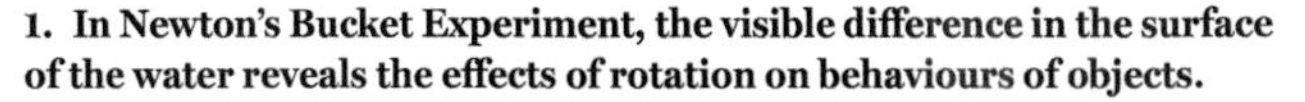

1. In Newton's Bucket Experiment, the visible difference in the surface of the water reveals the effects of rotation on behaviours of objects.

thing that distinguishes the first situation from the second—and hence it seems the only thing that can explain the difference in the shape of the water's surface—is that the bucket in the second case is rotating. This was meant to show that relations to space itself, in the form of differences in an object's state of motion, made an observable difference to its behaviour. The laws that predict that behaviour would have to refer to the bucket's state of motion. Newton can say that an accelerating object is changing its position in space and the water in the bucket is responding to that change. Leibniz, it would seem, has no explanation for the difference in the shape of the water's surface. There's nothing in the spatial relations between the parts of the bucket and the water inside that would account for why it rises in the one case and stays flat in the other. There is no term that he could put into his equations, so long as he confines himself to the spatial relations among things internal to the water and bucket system that would predict that effect.

That leaves open a loophole, however, that Newton anticipated and closed with his next example. While it is true that there is nothing in the internal spatial relations among the parts of the water and bucket system—aside from the shape of the water itself—that distinguishes the twirling bucket from the one at rest and could be used to explain the difference in the shape of the water, that is no longer true if we widen the description to include objects in the environment. If you do the experiment in your kitchen, for example, the twirling bucket will be changing its relationship to your fridge while the stationary one is not. Leibniz could in principle say that the water is responding to its changing relation to your fridge, rather than that it is changing relation to space. The experiment itself doesn't rule that out. Of course, you wouldn't want to give your fridge such a position of importance in the fundamental laws of nature, but there are objects like the centre of mass of the universe, for example, that could plausibly be thought to fill that role.

Newton closed the loophole by considering a fictional universe obeying his laws of motion that contains just two globes.

> If two globes, kept at a given distance one from the other by means of a cord that connects them, were revolved about their common center of gravity, we might, from the tension of the cord, discover the endeavor of the globes to recede from the axis of their motion, and from thence we might compute the quantity of their circular motions.... And thus we might find both the quantity and the determination of this circular motion, even in an immense vacuum, where there was nothing external or sensible with which the globes could be compared.

The observable tension in the cord, which pulls the globes towards the centre of motion, testifies to the absolute rotation of the spheres even when there is no relative motion in the entire universe. All material bodies (the globes and the cord) maintain a constant position with respect to one another. That means that as long as Newton's laws are correct, acceleration is detectable. Leibniz did not seem inclined to challenge Newton's laws, and that would have been a hard argument to make since there was nothing to suggest that those laws were anything but perfectly accurate. He never answered the bucket and globe arguments convincingly and the question was left hanging until much later.

The correspondence between Leibniz and Clarke contained the seeds of a form of argument that runs through the history of spacetime physics. The original argument between Newton and Leibniz concerned whether space is a thing in its own right or just a network of relations among bodies, but as part of his argument against recognizing space as a thing, Leibniz pointed out that such a view committed one to recognizing distinctions between situations that were indistinguishable from one another. He advocated a view that excised the unobservable structure.

That kind of argument appears again and again in physics and has been one of the most powerful tools leading us to new theories. In modern terms, we would say that the importance of the debate was that it led to a discussion about the appropriate physical geometry for dynamics. If one thinks space and time are both independently existing entities, as Newton did, there are differences in location and orientation *in space*, differences in uniform motion, and differences in location and duration *in time*, that are undetectable. It's not *just* that the theory postulates unobservable structure to explain observable behaviour. It is that Newton's theory postulates unobservable structure that plays no role supporting the observable behaviours of things; structure that can't be detected by any experiment and makes no difference at all to the law-governed movements of objects.

The particular question that interested Newton and Leibniz—that is, the question of whether space and time are things or networks of relations—has mostly fallen by the wayside. People nowadays speak of space and time without worrying too much about whether they are substances or a network of relations. The question that has survived and has been at the centre of theoretical developments since the debate between Newton and Leibniz is: what is the structure of space and time? The debate turned out to be very fruitful in identifying undetectable structure. And it introduced a way of thinking about these things, in terms of what sorts of transformations we could make to a system of objects without disturbing anything that would make an observable difference in their movements. The question 'What intrinsic structures in space and time need to be recognized to support the law-governed movements of objects?' became the central problem of spacetime physics.

The mathematical tools that Leibniz was employing implicitly by asking what kinds of shifts or changes one could make to a system of objects without detection were developed explicitly and have been used to get an ever better understanding of the intrinsic geometry of space and time. They were refined and honed over the years, formalized in the mathematical machinery of symmetry and transformations that turned out to be central to the development of relativity theory and have played an indispensable role since. There can be a lot of mathematical complexity in the deployment of that machinery but the basic idea is clear and intuitive. Space and time are not themselves observable. And if there are certain kinds of imagined transformations that one could make—either to the world as a whole, or to a system of objects in the world—without detection, we don't have a good reason for believing that those transformations change anything physically real. Insofar as we only know the structure of space and time by way of its effects on the motions of objects, we have no evidence for the existence of structure that makes no difference to observable motions.

Suppose somebody says to you, 'here are all of the ways the world could be, according to the physical laws', and for any way the world could be, you could obtain another way the world could be (according to the physical laws) by taking that one and changing it in this way. If the change always yields an indistinguishable counterpart, you should wonder whether these 'changes' are changes in name only. You should wonder, that is to say, whether the situations obtained by implementing the change in question really, after all, represent the very same kind of physical situation. Actual examples are complex in ways that require careful analysis, so this wondering is just an invitation to investigate further. What is clear is that whatever is transformed is not something that is observable or makes any known difference to the observable movements of objects and if we don't have an independent reason for thinking that the transformation changes something physically real, we would be right to question whether this is an empty change.

It is not always easy to see how to remove the undetectable structure in a way that makes physical sense. Unobservable structure can be so entwined with structure that plays a role in production of observable physical behaviour that the two cannot be easily separated. The example of velocity and acceleration provides a good example. Even though motion at a constant velocity is not detectable, acceleration is, and it is very unclear how to make physical sense of the idea that there is a fact about whether something is accelerating, but not a fact about whether it is moving at some particular velocity. That is not uncommon. It is typically clear only in retrospect how to excise undetectable structure while leaving everything in place in order to produce observable behaviour. It is typically only in retrospect, that is to say, that it becomes clear how to get rid of the fat without cutting into bone.

A quick recap of the ground we covered in this chapter: Newton brought questions about the intrinsic structure of space and time into physics by connecting them to motion. This changed the way that space and time are studied by making them subject to empirical investigation. Newton himself held that space and time were entities in which objects and events were located. Leibniz objected on the grounds that such a view committed Newton to an infinite number of distinct physical situations between which we have no way of distinguishing either by observation or experiment. The debate introduced the question that has been at the forefront of the physical investigation of space and time since Newton: what kind of structure do space and time have, as judged by their effects on the motion of objects? Or, as we would put it nowadays, what is the geometry of space and time, as gauged by their effects on the observable movements of material bodies? And the mathematical tools that were forged to make precise the way of arguing that appeared in the early debate between Leibniz and Clark have played a central role in the physical investigation of space and time since.

The story as I have told it in this chapter is not just the story of a new worldview, it is also the story of the birth of physics as we know it, that is, as a distinctive enterprise that branched off from the philosophical tradition. At this stage, the metaphysical debate between Newton and Leibniz is stalemated. What happens next is something no one could have anticipated.

Chapter 2
From space and time to spacetime: the era of Einstein

By the end of the 17th century, Newton's theory had gained complete acceptance, but there were some clouds on the horizon. There were clues that the theory wasn't entirely correct. These were at first little anomalies that couldn't be fitted into the existing framework but they grew increasingly pressing and Newton's theory would eventually be superseded by Einstein's theories of relativity. This chapter introduces the brilliant imaginative leaps of Einstein and the surprising new direction in which they led physics.

Let's start with the problem to which special relativity was a solution. According to Newton's laws, the law-governed movements of physical objects aren't sensitive to differences in where they are located, to what direction they are facing, or to differences in state of motion so long as that motion is uniform. Things behave in every detectable way the same no matter the location, orientation, or velocity. They are, however, sensitive to acceleration. Although it remained philosophically unresolved what to make of the fact that behaviour is sensitive to acceleration, the idea that the laws of physics are insensitive to differences in velocity had become deeply entrenched and had attained an almost sacrosanct status. In the 19th century, however, investigation into the phenomena of electricity, magnetism, and light culminated in Maxwell's equations of electromagnetism.

Maxwell's equations *did* seem to show sensitivity to differences in velocity. The equations entail that when light is generated, the electromagnetic waves go out in all directions at a speed of 2.998×10^8 m/s (186,000 miles/s) (the speed plays such an important role, it has its own label: it is customarily called c). They also entail that it doesn't matter if the source of the light is moving. In that respect, light is like sound. The speed of sound waves is independent of the motion of the source so that if you yell out of the window of a moving car, the sound moves no faster than if you had been standing still. This means that you should in principle be able to tell how fast you are moving in space by measuring the speed of light relative to you. If you are riding in a car that is going at 100 km/h, and a car speeds past you in the same direction going at speed 200 km/h, then the speed of the passing car relative to you should be (200–100) km/h. So it seems that if you are driving away from a lighthouse at velocity u and the light is travelling from the lighthouse at c, the velocity of light relative to you should be $c–u$, so by measuring the speed of the light from your perspective in the car, you should be able to determine how fast the car is going.

All of this was quite contrary to the presumption that there is no way to detect whether an object is at rest or what velocity it is moving at so long as that velocity is uniform, and a bunch of experiments in the 1800s tried to use this idea to see if they couldprove that presumption wrong. The most beautiful and famous of these was an experiment first performed by the physicist A. A. Michelson in Germany in 1880–1, later improved and performed in the USA together with the American chemist Edward Morley. It is known as the Michelson–Morley experiment, and here's how it works. Since the Earth is moving very fast around the Sun, we should be able to gauge its motion by seeing how the apparent speed of light differed from c when it was measured first in one direction and then in a direction perpendicular to that. The problem was that when Michelson and Morley performed the measurements, they found that it was exactly the same in every direction. Many different ways of trying to measure the speed were tried and the measurements were made with extreme delicacy, yet still the results came back the same. It became increasingly clear that it doesn't matter how fast and in what direction you move, you get the same result for measurements of the speed of light. It is extremely difficult to understand how to make sense of these results. People were stunned; nobody at the time understood how an observer moving very fast in one direction would get the same result as someone moving equally fast in the opposite direction.

Something like this happens often in science, usually in moments before important transitions. It is not that different from what happens with detectives developing a theory of a crime. One starts with a theory that is simple and plausible and seems to account for everything one knows in a natural way. Then a new piece of evidence is discovered that doesn't fit: maybe a footprint or a glove, or an unexpected cell tower ding. People try various ways of massaging the old theory to accommodate the recalcitrant evidence. They nudge this and add that, but nothing seems quite right. There doesn't seem any way of accounting for the new evidence that doesn't seem ad hoc, arbitrary, or made up. And then someone comes along with a completely new theory, one that reorganizes all of the old evidence together with the new and everything snaps into place. That is what happened here. The first thought that scientists had was that the trouble must lie in the equations of electromagnetism, since the edifice of Newtonian mechanics seemed inviolable and Maxwell's equations were quite new. But in time it became evident that those equations were correct and the trouble had to be elsewhere.

The Dutch physicist Hendrik Lorentz made a seemingly bizarre proposal. He suggested that the true speed of light for objects that are really at rest is c. For observers moving at a velocity v, the speed of light is $c-v$, but when they try to carry out the measurement, they get the result c because their measuring instruments trick them. The theory was that measuring rods shrink and clocks run slower when one is in motion in precisely such a way as to ensure that measurements of the speed of light never come out as anything but *c*. The terms 'measuring rods' and 'clocks' are being used generically here. Since velocity is distance per unit time, one is going to need some way of measuring distance and some way of measuring duration, and these measurements are going to have to use some physical standard of length and time. If we fool with those in the right way, we will get the right measured result across the board. In its own perverse way, the suggestion has a logic about it. It's crazy, of course, but at the time nobody had an alternative that seemed more sane.

Nobody, that is, until Einstein. It is unclear whether Einstein knew of the Michelson–Morley experiment. There are reasons to think he did and reasons to think he didn't. He was, however, steeped in the new electromagnetic theory and he had been trying unsuccessfully to modify Maxwell's equations so that they no longer entailed that the speed of light relative to observers moving at different velocities should be different. At some point, he changed tack and asked instead what the world would have to be like for the speed of light to be actually the same for every uniformly moving observer.

Before we can say what he found, we need to introduce a way of representing space and time that will be indispensable in what follows. Suppose we have a surface like a square piece of paper sitting on a table. There is a spatial distance between any two points on the paper. If we now let the vertical dimension above the table represent time, the result will be a three-dimensional volume, two dimensions of which represent space and one time. Each of the points inside the volume will represent a position at a particular time. The history of an object like an ant crawling around over the surface of the paper when we plot its position at different times in the cube will be a line that rises continuously up

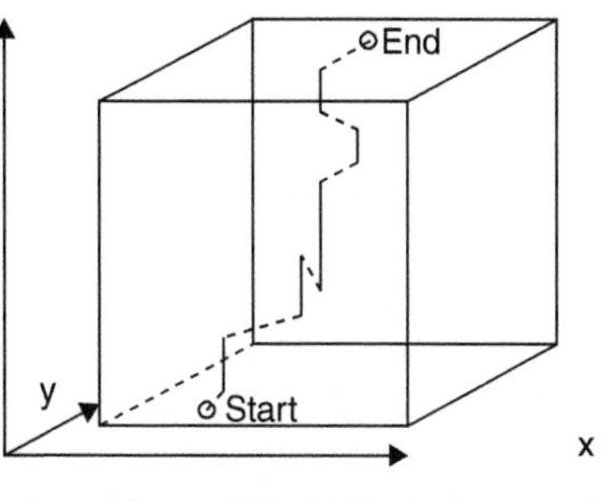

2. A spacetime cube can be used to plot the trajectory of an ant over a piece of paper lying on a table by utilizing the vertical dimension to represent time.

the vertical dimension as it traces a path through the other two dimensions that reflect its movements in space (Figure 2).

Lines in spacetime representing the histories of objects like this are called 'trajectories'. If you take a run through your neighbourhood following the path depicted in Figure 3(a), your trajectory in spacetime would look like the line in Figure 3(b).

Representing your trajectory in this format elevates your path through space so that we can locate your position at a given time by finding the point corresponding to that time along the vertical dimension.

Our space has three dimensions rather than two, so the spacetime of our world has four dimensions rather than three. Otherwise it is the same deal. Any event has a location in both space and time. And just as we ordinarily think that a straight line between any two points in space gives the distance between them and a straight line between any two events gives the temporal interval separating them, a straight line between any two events in spacetime corresponds to the spatiotemporal distance between them. So, for example, since Galileo was born on 15 February 1564 in Pisa, and Newton was born on 4 January 1643 in Woolsthorpe, the

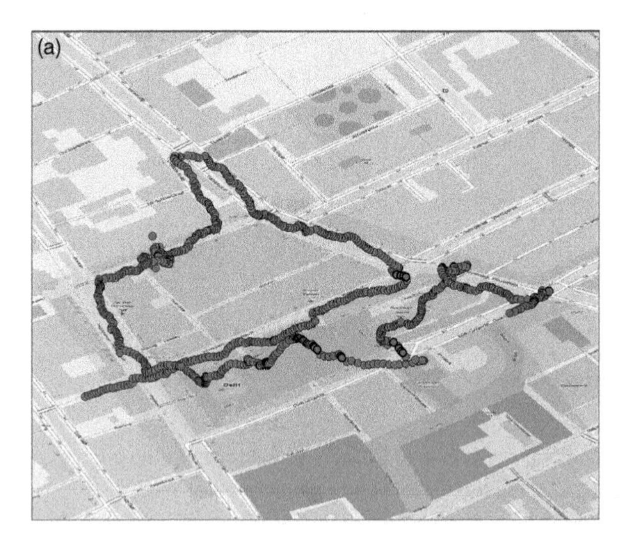

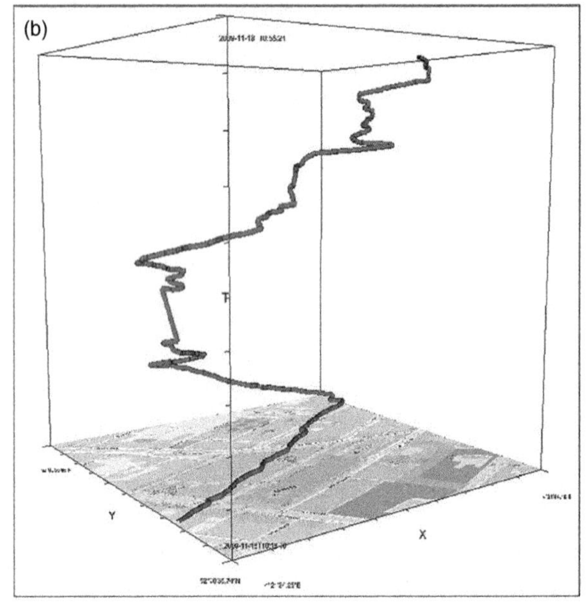

3. This pair of figures shows the same path plotted in space (a) and spacetime (b).

spatiotemporal distance between their births is represented by the straightest line connecting them in spacetime. We could measure that distance by calculating the amount of time it would take a light ray to get from one event to the other.

Now we are in a position to see what happens if we start from the assumption that the speed of light really is the same for any uniformly moving observer and draw out the implications. It turns out that spatiotemporal distances between events don't vary, but observers moving at different velocities will split the difference between space and time in different ways, so they get different results for spatial distances and temporal distances. The difference is so subtle that it is only noticeable at very great distances or speeds close to *c*. Lorentz had already provided the equations that let us calculate the difference, but where he thought of the equations as describing the ways that rods shrink and clocks slow when in motion, Einstein embraced it as revealing that there really aren't objective facts about spatial distances and temporal intervals. Once we make that adjustment, we can combine Newtonian mechanics and electromagnetism and account for the fact that observed velocity of light is independent of the motion of the observer. The philosophical impact of the adjustment on our understanding of time, however, is profound.

At the time that Einstein was working, people were still using reference frames to think about the structure of space. They laid down grids with numbers that served as coordinates and spoke of transformations between these frames of reference. An event was represented by four numbers, three of which gave its spatial coordinates and one gave its time. The nice thing about this way of representing things was that it allowed questions of geometry to be converted into arithmetical problems, and we can use mathematics to solve equations. If we introduced reference frames centred on observers in different locations or moving with respect to one another, we could find equations that would tell us how their measurements would relate to one another. The not-so-nice thing is that it made the physical content of our equations obscure. Recall that I said in Chapter 1 that it is convenient to use numbers as names for locations, namely because numbers have structure we can exploit to represent structure among the places we use them to represent.

Only some features of the numbers that we use typically have this kind of physical significance. So, for example, we usually choose an origin for our frame of reference and a scale by convention. These are choices that are arbitrary in the sense that we could have chosen a different origin or scale to represent the same set of physical facts. This doesn't present a problem so long as we know how to separate the features of our coordinates that carry physical significance from those that don't. When we are coordinatizing a space whose intrinsic structure is unknown, it is not trivial to distinguish features of coordinates that are representing something physical from those that are artefacts of the choice of reference frame. Even though we had the equations that told us how the measurements of one observer would relate to the measurements of another, it wasn't at all clear by looking at the equations what all of the numbers were telling us about the structures intrinsic to space and time.

If you understand that the Earth is round, you understand why people standing on opposite sides of the Earth disagree about which direction is up. One would like something similar in this case. One would like a clear understanding of the intrinsic geometry of the universe so that we could understand why measurements of spatial and temporal distances depend in the predicted ways on our state of motion. Hermann Minkowski's presentation in terms of Einstein's theory offered that. Minkowski was Einstein's mathematics professor at Zürich Polytechnic. Einstein attended a few of his lectures, but later admitted that he was so absorbed in his study of physics that he neglected his classes in mathematics. It was Minkowski who supplied the four-dimensional formalism that allows us to represent physical geometry directly. In this formalism, coordinates disappear from the picture. Space and time are presented together using the kind of four-dimensional construction described above, and instead of working with numbers, we simply talk about the intrinsic geometry of this four-dimensional structure. This lets us present the physical ideas geometrically and makes it much easier to separate what is part of the actual structure of spacetime from things that are artefacts of arbitrary choices for assigning coordinates. We will see this more clearly in Chapter 3 when we see why spatial and temporal distances seem to vary with the motion of the observer.

Einstein resisted the formalism at first, but it was embraced almost immediately by the rest of the physics community. It quickly became the canonical way of presenting it, and Einstein eventually came around. It was once said that nobody really thinks in four dimensions but most physicists now think easily and naturally in terms of spacetime.

If you consider a pulse of light emitted from a point p in spacetime, it will spread out in every direction at the same speed, regardless of the motion of the source. The expanding sphere of light appears in a spacetime diagram as a cone, known as a light cone (Figure 4).

This is one of the areas in physics in which the terminology is mnemonic. The interior of the cone in the future direction (the future light cone) contains all of the events that can be affected by p. The interior of the cone in the past direction (the past light cone) contains all of the events that can affect p. The region outside the light cone on both sides—called, delightfully, the 'absolute elsewhere'—contains events that can neither affect nor be affected by what happens at p. All that you need to know about the structure of spacetime according to special relativity is that there is a light cone at every point that separates the other points in the space into these three classes. In practical terms, 'past' and

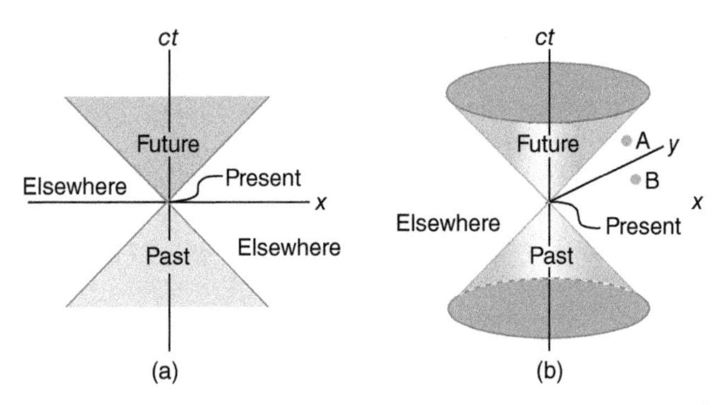

4. Light cone diagrams are used to represent the intrinsic structure of spacetime in a relativistic theory. The expanding sphere of light released from a point will define a cone whose interior in the past and future directions, respectively, represent the set of points that can affect and be affected by what happens at the point.

'future' have meanings quite close to their everyday meaning. A is in B's past means that A can affect what happens at B. C is in D's future means that C can be affected by what happens at D. An event's past is now given by the contents of its past light cone; its future is given by the events in its future light cone. The difference is that there is now also a great grey area containing events that happen neither before nor after the event in question, nor (properly speaking) at the same time.

The transformation of the imagination was made possible by the introduction of the geometrical way of thinking and the sidelining of the clumsy use of coordinate systems. As is often the case in physics, the honing of the formalism and clarification of its physical content happened in tandem. With the new formalism we can understand the real significance of Einstein's innovation. The trick was to combine space and time into a single structure, put aside any presumptions about the differences between the two, and recognize only those structures that are manifested in the law-governed behaviours of physical objects. It turns out that the real objective structure of spacetime is defined by spatiotemporal distances and the constancy of the speed of light is a natural consequence of the fact that it spreads out evenly from a source. This is the gold standard in physical explanation. You want to start with a collection of behaviours that seem bizarre by the lights of your existing understanding of the world, and end with new fundamental concepts and a new mathematical framework from which they flow exactly as you would expect. There is an immediately manifest kind of naturalness that we have learned to expect from a physical framework that has any chance of being correct: an economy of fundamental concepts and an inner elegance. Einstein's theory had those in spades. But once we have the new framework, we are left to assimilate it philosophically and Einstein's innovations did such violence to fundamental pieces of common sense that, as we will see in chapters to come, we are still coping with the fallout.

How all of this changes our philosophical understanding of time

Einstein's theory revived an ancient philosophical debate between the Greek philosophers Heraclitus and Parmenides. You may know Heraclitus from the claim that 'all is flux' or the observation that one never steps into the same river twice. We only have fragments of his writings. The fragments are more like aphorisms than theories, but they are evocative enough that you get the flavour of his ideas. For Heraclitus the fundamental character of reality is *change*. As he puts it:

> Everything flows and nothing abides; everything gives way and nothing stays fixed...
>
> Transience is basic, and the present is primary. Those things which exist now do not abide. They slip into the past and non-existence, devoured by time.

The opposing vision of time comes from Parmenides. According to Parmenides, the universe is constant, unchanging, and eternal. He writes:

> What Is has no beginning and never will be destroyed: it is whole, still, and without end. It neither was nor will be, it simply is—now, altogether, one, continuous..."
>
> Permanence is basic. No things come to be or, slipping into the past, cease to be. Past, present, and future are distinctions not marked in the static Is. Time and becoming are at best secondary, at worst illusory.

To many in the physics community, the special theory of relativity seemed a vindication of the Parmenidean vision. It became customary to say that the passage of time is an illusion and that nothing really changes. Einstein himself used the vocabulary of illusion on several occasions, most famously in a condolence letter to the widow of his close friend Michele Besso upon learning of his death. He wrote: 'In quitting this strange world he has once again preceded me by just a little. That doesn't mean anything. For those of us who believe in physics, this distinction between past, present, and future is only an illusion, however persistent.'

We will dig a little more deeply into the debate between these two visions of time in Chapter 5, but there are some simple confusions that we should clear up right away. By itself, the fact that we can represent time mathematically by a dimension means very little. It certainly doesn't mean that time is just like space. We can represent temperature, saturation, and hue by dimensions and there is little temptation in these cases to say they are just like space. Any time you see a graph, someone is using a dimension to represent a quantity. We could take the spacetime cube in Figure 3, and instead of letting the vertical dimension represent time, let it represent the level of infection, the amount of crime, or the distribution of wealth across the space pictured on the map below. When the relativistic image of time is introduced, it is sometimes said that according to relativity the future is already there, or that according to relativity, there's no such thing as change. Neither of these things is true. It is not true that the future is *already* there, or there *now*. In a relativistic world, things happen when they happen and at no other time, just as they do in a Newtonian world, and just as they do in life. Nor is it true that nothing changes. In a relativistic world, trees grow; leaves change their colour; the tides ebb and flow. The fact that the 4D image isn't changing does not mean that it is not an image *of* change.

There are, however, features of common sense that are difficult to reconcile with special relativity. Common sense tends to think of the universe as a large spatially extended thing that unfolds in time. And the differences between time and space are intuitively quite deep. Space has a kind of substantial reality; it is where material objects are housed even when we are not looking at them. Time seems to have a less substantial existence. We tend to think that other times aren't like places that exist even when they are not in view, but rather that they come into existence *as they are experienced*. These common-sense ways of thinking are very hard to maintain in a special relativistic universe. The universe doesn't separate into what we might call 'same-time slices' that come into being one at a time and then go out of existence again. That means that you can't accurately think of time in a relativistic universe as a kind of external parameter in which the history of the universe unfolds. From within the universe, individual temporal processes look more or less the same. Our own lives, and the histories of other material objects, pass through a sequence of states that occur one at a time, and the normal spatial and temporal fabric of everyday life remains largely intact. It's just that there's an asynchrony between processes that are unfolding at a spatiotemporal distance from one another, so that if we look very closely the fabric of spacetime is more like a mesh than a neat slicing into three-dimensional spatial surfaces. If we normally think the universe looks as per Figure 5, where each of the sheets

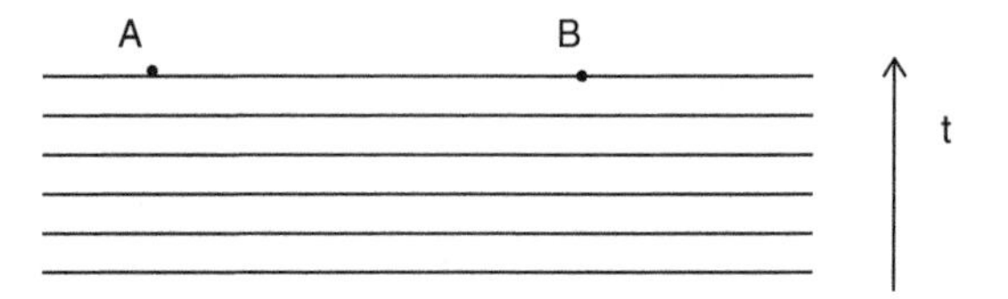

5. Pre-relativistic common sense assumes that spacetime divides into slices representing events that occur at successive instants of time. For any two events, there is a determinate fact about whether they occur at the same time.

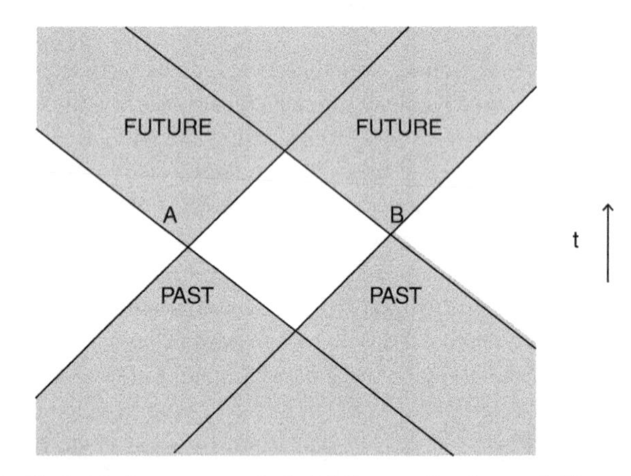

6. The intrinsic structure of relativistic spacetime is a mesh-like structure defined by light cones in which there is no well-defined fact about whether distant events like A and B that don't fall within one another's light cones happen at the same time.

represents what happens at an instant of time, relativity tells us instead that it looks as per Figure 6.

When we talk about the geometry of spacetime, we've stripped things down to the most austere structures. At this basic level, space and time encode relations of causal connectability among events. How things are located in space and time tells us whether, and by what route, they can affect one another. Something over here can only affect something over there by a signal or sequence of influence that passes through the space between. Something that happened in 1903 can only affect events in 1925, or be affected by events in 1880, by a route that passes through all the times between. The structure of space and time together encode the flow of influence in the world. What special relativity did, by placing an upper bound on how fast anything can travel from one part of spacetime to another, was place an upper bound on how tight the causal ordering is.

Suppose you and I are in ships on the open water and that we communicate by signals that we send back and forth. We each know what is happening along our own timelines, but if I want to relate the events along your timeline to mine, I have to send you a signal that travels from my ship to yours, asking what you are doing at the precise moment you receive the message and wait to hear back. If you tell me you were sitting down to dinner when you received the message, then I know you sat down to dinner between the time I sent the signal and heard back, but I can't narrow it down any more than that. A natural tactic that will have occurred to you is that I should just divide the time it took the signal to come back to me by two; that should tell me that you were sitting to dinner exactly halfway between when the signal was sent and received. But if we think this through more carefully in the world of special relativity, we will see that it won't work.

Dividing by two makes sense if we are both moving at the same velocity, but if you are moving at some constant velocity away from me, then, from my perspective, it will take a shorter time for the signal to get to you than to return. We can calculate, given the speed and distance, what the correction for the different length of the return trip should be, so that's not a problem. But now let's see how your calculations will go. You will be using the same techniques to coordinate events along my timeline with events along yours and making the same correction in the opposite direction because you will be treating yourself as stationary and me as moving away from you. From your perspective, it should take any signal you send less time to get to me than to return, and when you make your corrections and we compare, we will find that we have got different results for which events on your timeline happened at the same time as those on mine. I will be saying that you sat down for dinner before I turned on the TV, for example, and you will say the opposite. Which of us is right?

The whole point of relativity is that there is no right or wrong to speak of here, since there is no fact about who is moving and who is at rest. That is what the 'relativity' in 'special relativity' amounts to. (We will see below where the 'special' comes from). I'm moving at velocity v relative to you and you are moving at velocity $-v$ relative to me, and there is no fact about which of us is really stationary and whose calculations are really correct. That, in its turn, means that there is no fact about which events along your timeline actually happened at the same time as which events on mine. The smaller we make the lag time between sending and receiving signals, the tighter we can make the intervals within which we can locate distant events on our own timeline, but we never eliminate it entirely. We never eliminate it entirely because nothing travels faster than light. Even if we use light to signal, there will always be lag time. The asynchrony, or looseness of fit, is so slight that—as I said above—it is unnoticeable at everyday speeds and distances but it becomes significant at larger scales. It is essential to getting the right predictions for the Michelson–Morley experiment and it is absolutely crucial in getting the sharp edges of our worldview to line up right.

This is a characteristic example of how physics works. There was a direct confrontation of a worldview with a very careful measurement that forced a deep and thorough rethinking of fundamentals. The worldview in this case was the broadly Newtonian one that combined the laws of electromagnetism with the Newtonian mechanical laws, and it was expressed with enough mathematical precision to entail a definite prediction for the result of an experiment designed to detect the impact of the motion of the Earth on the measured speed of light. The result that was actually obtained was not what was predicted, and that meant something was wrong. A new theory was put forward to accommodate the phenomena, and a deeply entrenched philosophical belief—one that is foundational to common sense—falls. It takes a generation to assimilate the impact. Philosophical ideas in physics are driven by this kind of pressure. One builds a philosophical system with enough mathematical precision that it can make predictions in an exact and definite way, and then lets the world tell us whether the prediction is correct.

The comparison between prediction and observation is the point of confrontation between theory and world. When there is a conflict, and we have assured ourselves that the observation is correct, some part of our theory has to go. The conflict itself won't tell us what to change and people typically try all kinds of things. Again and again, in retrospect, the ideas that end up working out are the ones that are ruthless to pre-scientific philosophical ideas but preserve the elegance of the mathematical apparatus.

Mathematical elegance is as difficult to characterize explicitly as elegance of design in engineering. It has to do with simplicity, symmetry, the absence of arbitrary, ad hoc, or inessential structure. The kinds of criteria that detectives invoke in judging between different theories of what happened to produce the evidence found at a complex crime scene are not far off the mark. You have a body in a windowless room, locked from the inside and some odd but seemingly irrelevant details: a stolen letter, a sick dog, a flower missing from a bouquet, and a safety pin attached to a drape. At first nobody knows what any of these things have to do with one another, but then someone comes along that fits it all together and everything snaps into place. That sense of things snapping into place is what one is after in constructing a theory. Just as Copernicus took a leap of the imagination, upsetting a deeply entrenched view of the universe to accommodate the movements of seven wayward points of light in the night sky, Einstein showed us how to put Newtonian mechanics and electromagnetism together in a way that accommodated the observed constancy of the speed of light, and the result was completely transformative of our understanding of time.

Reconstruction

The familiar world of everyday experience has to be deconstructed and put back together on this foundation. Space and time form a four-dimensional structure with a mesh-like character. Material bodies—tables and chairs and persons and penguins—are as real and solid as ever, but they appear in this framework as dense clusters of material events that run along the temporal dimension. It remains true in a relativistic world that stars are formed and mountains emerge, forests grow, and fish die. Mosquitoes and mole-rats and whole generations of people live out their lives as they always did. We keep track of what we see and use the events in our own lives as landmarks, or points of reference, ordering other events by their relationship to the here and now. This is relatively simple for the individual observer who thinks of the events she sees and hears as unfolding with her perceptions of them. It is a little more complex with a collection of individuals getting information from a range of places and sharing it with one another.

The way that communication and coordination get organized into universal time with the dating systems as described in Chapter 1, and from there into Newton's absolute space and time, is natural but not inevitable. It lets us coordinate the time of events across distances at the level of precision that we need for everyday communication. It tells us what is past and done with, and unaffected by anything we do in the here and now. It tells us what remains to be done, how to get there, and what to do to influence what happens at other places and times. Special relativity does the same thing, but it provides us with a much deeper understanding of the basic relationships.

General relativity

We are not quite finished with Einstein's innovations. Special relativity is called 'special' because while it provides a good account of the spatiotemporal framework needed for the laws that govern how things move and the phenomena associated with electricity and magnetism, it wasn't compatible with gravity, at least as it was understood by Newton. Newton's laws said that gravity is a force of attraction that acts instantaneously and depends on the distance between two bodies. In special relativity, however, we saw there are no objective spatial and temporal distances. It no longer makes sense to talk about which distant events happen at the *same time*.

A simple way to fix the problem would have been to reformulate the law of gravitation in terms of *spatiotemporal* distances, and a good many of Einstein's colleagues adopted that strategy. Einstein thought the root of the problem was deeper and he wanted to get to the bottom of it. Because his breakthroughs were so dramatic, there is a good deal written about Einstein's methods. As a human being, he was a complicated and charismatic figure: a humanist as much as a scientist, and a philosopher down to his bones. Indeed, it was a philosophical conviction that partly motivated the search for the general theory of relativity. He thought that there was a deep flaw in a theory that had to recognize a fundamental distinction between bodies that were accelerating and those that were moving at a constant velocity. He sided with Leibniz in the debate with Newton and was bothered by that old Newtonian distinction whose necessity was illustrated by the bucket and globes. It is widely agreed that Einstein's new theory didn't live up to his philosophical aspirations to provide a fully relational theory of space and time, or at least that the story is a little more equivocal and less clear than he had hoped. But no matter, the theory that resulted is beautiful. Most physicists will tell you it is one of the most beautiful theories ever invented. It remains one of the pillars of modern physics.

Einstein's life is so intertwined with the history of physics in the 20th century that it is inconceivable without him. This was an unparalleled period of triumphant breakthroughs, and he was at the centre of the most important of these. Einstein's early work came fast and furious. He was only 21 (famously working in a patent office because he was unable to find a position in a university) when he published the paper that introduced the special theory of relativity. That same year, he published three other papers on separate topics, each of which had a profound effect on the development of physics. The struggle with gravity was, by contrast, a grind. He worked for eight long years to incorporate gravity into the relativistic framework. The struggle began in 1907 with what he later called 'the happiest thought of his life' and ended in 1915 with four lectures in Göttingen that culminated with the presentation of the general theory of relativity.

The happy thought that led to the theory was the observation that from the inside, being in a gravitational field is just like accelerating in deep space away from any source of gravity and (conversely) being in an elevator falling freely towards the Earth is just like being in deep space away from any gravitational field. If you were in an elevator out in deep space being propelled upward, balls released from hands would fall downward, just as they do on the surface of the Earth. You could have a whole lab there with you, so that you could carry out whatever experiment you want, and you wouldn't be able to tell whether the lab was in a building on Earth or accelerating upward in a rocket at 9.81 m/s^2.

And likewise, if you were in an elevator in free fall from a very great height, everything would behave just as it does for astronauts in outer space. You would float above the floor of the elevator and pass bowling balls back and forth like they were balloons.

This should remind you of Galileo's ship. You may recall the argument that there is no experiment that someone on a ship moving at constant velocity could carry out that would have a result different from someone on shore. Likewise, there is no experiment that someone in an elevator could carry out that would tell them whether they were accelerating upward in deep space or sitting still in the presence of a gravitational field. Again, what we see here is that if we use the law-governed movements of objects as a gauge, there is no difference between being at rest in a gravitational field and accelerating in a direction opposite to that of the field. And just as Leibniz proposed that we eliminate the difference between indistinguishable-from-the-inside worlds in which the whole network of material objects is moved while maintaining their relative positions, Einstein proposed that we eliminate the difference between these situations. He formulated a principle that he called the Equivalence Principle which says that there is no physical difference between uniform acceleration and the presence of a uniform gravitational field and searched for a theoretical framework that subsumed the special theory of relativity and satisfies the Equivalence Principle. He wasn't just motivated by philosophical conviction. There were other clues that motivated his search. One of the most important was something that had puzzled people for some time: in Newtonian physics, the term for mass in Newton's second law always takes the same value as the term for mass in the law of universal gravitation. This looks like a weird coincidence, because there is nothing in Newton's theory that suggests these have to be the same. If the Equivalence Principle is correct, however, inertial and gravitational mass are the same thing and this is just what you'd expect.

It took years for Einstein to find the theoretical framework he was looking for, and the key turned out to lie in mathematical tools that were discovered in the 19th century. Consider a lineage of

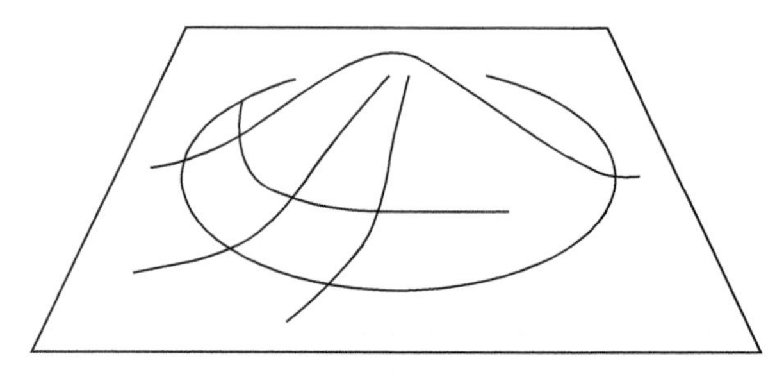

7. Gauss discovered that the geometry of curved spatial surface is non-Euclidean. His discovery led to the theory of curved spaces in higher dimensions that provided the mathematical basis for general relativity.

two-dimensional ants whose lives and measuring instruments and observations are confined to a two-dimensional surface. Suppose that this surface is not flat but has a bump in it as shown in Figure 7.

Even though the ants are themselves two dimensional and confined to the surface, if they knew Euclidean geometry, they could tell that their space is curved. Euclid's geometry says that the circumference of a circle should be 2π multiplied by its radius. If the ants measure the circumference of the circle around the bump and divide by the length of the lines leading from the circle's edge to the centre, they would find that it is less than 2π and they could infer from that that their surface must be raised in a third dimension. The German mathematician Carl Gauss who made this discovery about two-dimensional spaces wondered if it was possible to have curved spaces in higher dimensions, and that led to the mathematical theory of curved space known today as Riemannian geometry.

The Equivalence Principle and the development of Riemannian geometry all came together in what turned into the general theory of relativity. Einstein found a set of equations that relate the curvature of spacetime to the distribution of matter. The equations are local, which means that they relate the curvature at a point (or in its immediate neighbourhood) to the matter at the point (or in its immediate neighbourhood). They are mathematically complex and difficult to solve, but conceptually the theory is crystalline in simplicity. According to Einstein's theory there is no such thing as gravity. All objects in space follow the straightest trajectories that are possible, but space itself is curved by the presence of matter so that objects following straight-line trajectories accelerate. As the physicist John Wheeler succinctly put it: 'Spacetime tells matter how to move; matter tells spacetime how to curve.'

You may recall that we ended Chapter 1 with a puzzle that was left hanging. Although Newton's laws didn't allow us to tell the difference between uniform motion and rest, they did allow us to tell the difference between uniform motion and acceleration. That is what Newton's examples of the twirling bucket and circulating globes showed. If we were on an accelerating ship, water would start rising up the sides of buckets and cords holding massive objects would show tension in a direction away from the direction of acceleration. Before Einstein, nobody knew how to recognize a difference between uniform motion and acceleration without also recognizing a distinction between rest and uniform motion. Special relativity showed us how. Einstein hoped that general relativity would also eliminate the distinction between acceleration and uniform motion. In actuality, the theory doesn't so much *eliminate* that distinction as absorb it into the curvature of spacetime.

In 1915 when Einstein announced the theory, it was all pencil and paper speculation. The theory did make some novel predictions, but they were difficult to test and it wasn't until 1919, when Arthur Eddington made an expedition to an island off the west coast of Africa to observe a solar eclipse, that one of the strangest and most unanticipated predictions of general relativity—that light rays would be bent by the gravitational force of the Sun, a phenomenon that we had absolutely no other reason to expect—was confirmed. That confirmation was one of the most spectacular successes of science and almost instantly propelled Einstein to world fame. Einstein was so confident in his theory on the strength of its beauty that he is said to have reacted nonchalantly to the news that his theory had been verified. When asked how he'd have reacted if it *hadn't* been vindicated, he replied: 'I would have felt sorry for the dear Lord. The theory is correct.'

He did not, however, embrace all of what we now know to be implications of general relativity. In 1939, he published a paper in the journal *Annals of Mathematics* called 'On a Stationary System with Spherical Symmetry Consisting of Many Gravitating Masses', in which he sought to show that black holes—celestial objects so dense that their gravity prevents even light from escaping—were impossible. Nowadays research into black holes is one of the most active areas in astrophysics, but the recognition that our own universe might contain them emerged only in stages in the decades following 1915.

When I said earlier that the equations of general relativity are mathematically complex and difficult to solve, I didn't say quite how difficult. They are represented by ten non-linear partial differential equations, and even today there are only a handful of known solutions. The first of these was found only a year after the publication of general relativity by the German physicist Karl Schwarzschild. One of the features of this mathematical solution is that for very compact, high-density stars, it becomes much harder to escape the gravitational field of the star. Eventually, there comes a point where every particle, even light, becomes gravitationally trapped. This point of no escape is called the event horizon.

Even though the event horizon played an integral part in Schwarzschild's solution, most physicists did not believe initially that black holes actually exist. There was no reason to think that the conditions under which such exotic objects could form were realistic. In 1939, the physicists Oppenheimer and Volkoff provided a mechanism for the formation of a black hole. They argued that if a neutron star grew too massive, it should collapse under its own weight to an infinitesimal point, leaving behind only its ultra-intense gravitational field, and their work foreshadowed astrophysicists' current understanding of stellar-mass black holes.

It wasn't until the 1960s, however, when Roger Penrose proved rigorously that, under a wide range of circumstances, the formation of a black hole was almost inevitable, that black holes were accepted into the repertoire of real astrophysical objects. There is now a wealth of evidence for the existence of black holes. Even though black holes are themselves invisible since they are regions of space from which no light emerges, they tend to capture gas around their margins due to their gravitational pull. This gas is compressed into a superheated rotating disk. These have been observed and become a focus of study for astronomers. Astronomers have found stars orbiting invisible companions. The detection of gravitational waves created by merging black holes in the LIGO experiments of 2015 was enough to convince even the most hardened sceptic.

The reason for talking about this here is that time starts to behave very strangely in the vicinity of a black hole. It is sometimes said that time stops at the horizon of a black hole. That is not quite correct. What is correct is that from the perspective of an outside observer watching an object travelling towards the black hole, time seems to slow almost to the point of stopping. Suppose that you are sitting outside the event horizon at a safe distance and you see a clock heading towards the horizon of the black hole. You might expect that you should be able to see the clock fall into the black hole and disappear. What you actually see is that as the clock approaches the horizon, it will tick slower and slower, and fall more and more slowly. The slowing that you observe is really an optical effect caused by the way that paths of light rays are bent near the horizon; the light travelling from the clock to you will be increasingly delayed as it gets closer to the horizon.

You never see the clock either reach the horizon or stop completely; it goes on forever, getting closer and closer, ticking slower and slower. If we follow the clock's actual trajectory, it will cross the horizon smoothly and arrive at the singularity—the point at the centre of the black hole where density and gravity become infinite and spacetime curves infinitely—in a finite time. This all sounds very strange from a common-sense point of view, but it is perfectly intelligible in the framework of general relativity, and it's another example of the strange power of physical theorizing. We formulate theories guided by the kinds of considerations that we described in the last two chapters and then we draw out implications of the theory, and this is where we end up: in a world with dark objects from which no light escapes and around which time appears to stop.

Scientific cosmology

General relativity also transformed the way that cosmology was practised. Before the 1900s, cosmology was thought of as the science that studies large-scale extraterrestrial systems such as galaxies. Two years after the publication of general relativity, Einstein published a paper that changed that. The paper was called 'Cosmological Considerations in the General Theory of Relativity' and it transformed cosmology into a science specifically devoted to the study of the universe as a whole. According to Newtonian mechanics and special relativity the universe was flat and infinite. It didn't depend on how matter is distributed or moved around. There wasn't much of interest to say about its overall shape, and you didn't need to know anything specific about our universe to say it. In general relativity, that is no longer true. The universe is curved and possibly finite. It has a structure or shape that is determined by how matter is distributed. That makes the overall structure of our universe an object of great interest, but also something that it is very difficult to know. The information that we have about the universe is drawn from a very small patch of it. Because light travels at a finite speed nothing in our part of the universe can carry information about anything that happened outside a spherical volume centred on the Earth 8.8×10^{26} m in diameter. This provides an insurmountable limit that, depending on its size, puts the vast majority of the universe beyond our reach. Even if we had complete information about the distribution of matter in the light accessible region of our universe, that would be akin to having a flashlight that illuminates a small patch of a room of unknown size.

That hasn't stopped cosmologists from speculating. The way around it is to make some quite strong assumptions on how what we see relates to what could be seen from other vantage points. The accidental discovery of the Cosmic Microwave Background in 1964 provided observational data that made cosmology one of the most active parts of physics. The Cosmic Microwave Background is a faint radiation filling all of space. This turned out to be a rich source of information about the accessible patch of the universe, and it has yielded what is known as the Standard Model of Cosmology. The Standard Model is more speculative than the physics that we've spoken about so far. It makes the audacious assumption that the universe looks the same on sufficiently large scales from any vantage point and extrapolates general relativity to scales fourteen orders of magnitude greater than that at which it has been tested. But it constitutes the best current understanding of what the universe looks like at large scales. It offers a sobering vision of our tumultuous little lives against the background of the black expanse of space.

According to the Standard Model, the universe is almost featureless at sufficiently large scales. Matter is distributed more or less uniformly and it looks the same in every direction. It is, moreover, expanding. If we take spatial cross-sections of the universe, choose any pair of points on one cross-section, and compare distances between those points on earlier and later slices, we find that they are moving away from one another like dots drawn on an expanding balloon. As the expansion is extrapolated into the past, the universe becomes increasingly hot and dense until a point around 14 billion years ago where the quantities in the equations become infinite and stop making physical sense. It is presumed that the reason that the equations stop making sense is that quantum effects (the effects of the submicroscopic behaviours of particles described by quantum mechanics, which can be ignored when working at larger scales) become important at that point. The rapid expansion that happens in the moments after is called the Big Bang. Since the model can't be extended farther back, the Big Bang marks the beginning of the known universe.

Cosmology lends itself to beautiful imagery and there is an understandable public interest in what our universe looks like at the large scale. Equations take a lot of deciphering, but if you are given an image, you have a sense that you understand what a theory says. Images such as Figure 8 are used to convey the content of the Standard Model.

One has to be careful, however. Images like this are misleading in a way that is directly relevant to the philosophical lessons of relativity. The diagram has a timeline along the bottom and lines of longitude carve the universe into spatial slices. This can give the impression that the diagram provides a bird's eye view of a universe that unfolds in time and the lines of longitude drawn across the whole mark stages in its development that it passes through one at a time. That would be a mistake. Those lines of longitude are there for convenience to allow us to take a cross-section of the manifold and compare the spatial geometry on these slices. There is no objective way of carving the universe into these same-time slices and no external time in which the history of the universe unfolds.

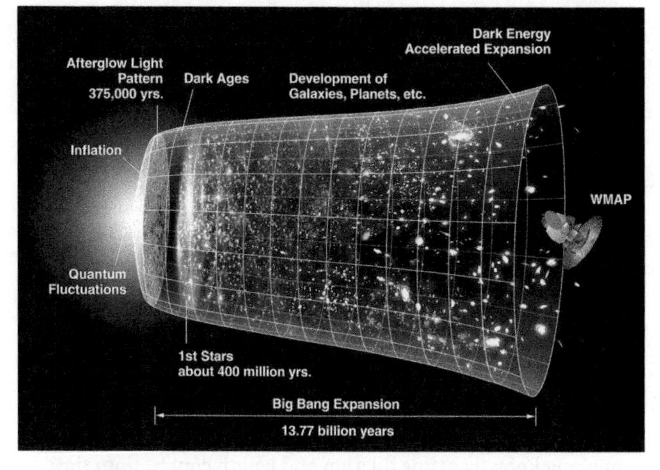

8. This diagram represents the Standard Model of Cosmology. Time runs horizontally across the bottom of the page; the bright spot on the left is the Big Bang and the increase in areas of the spatial cross-sections represents the expansion of space.

The universe is a four-dimensional network with a light cone at each point that can be extended to form a mesh-like structure consisting of processes that unfold asynchronously. There just is no fact about whether distant events happen at the same time. If you want to imagine the universe as a whole, you have to leave time inside the frame. It takes some time to get used to this way of thinking and we'll be exploring its implications in chapters to come.

Chapter 3
Philosophical implications of relativity

In this chapter, we look at some of the counterintuitive implications of the special theory of relativity. We start with the phenomena known as time dilation and length contraction: the fact that measurements of spatial distances and temporal intervals vary with the motion of the observer. We already saw a glimmer of this in Chapter 2 when I spoke of the ships trying to coordinate events along their own timelines. The intrinsic structure of spacetime is fixed and is given by spatiotemporal distances between points. Nothing is actually varying; time is not speeding up or slowing down. Clocks are not running fast or slow. Measuring rods are not shrinking or expanding. What is actually happening is that people at motion with respect to one another and using a certain convention to assign times to events end up disagreeing with one another about which sets of events are happening at the *same* time, rather like people on opposite sides of the Earth using the same convention (up = in the direction of the sky) are going to end up disagreeing which direction is up. Disagreements in which events are happening at the *same* time will translate into disagreements about temporal intervals and spatial distances; everything gets roundly messed up. We can get rid of these weird and artificial effects by sticking to the intrinsic geometry. That means talking about the spatiotemporal distances between events and not carving them up artificially into distances and durations.

Let's see how this works more precisely. Suppose that I have a watch and you have a watch. If you and I employ what seems like the most natural way for assigning temporal coordinates, it will turn out that if we are moving at a constant velocity (speed + direction) relative to one another, we are going to assign different times to distant events. Here is how we assign times to events. I have a team of people, each with their own watches and at rest relative to me, distributed at convenient locations around space: some close by, some far away. I synchronize my watch with theirs by dividing the time it takes for a light signal to travel to them and back by two. You do the same. You have your own watch and your own team. Your team is spread out evenly across space and at rest relative to you. You synchronize your watch with the watches of your team members by dividing the time it takes for a signal to travel back and forth between you by two. I trust my team, you trust yours, and we each carve spacetime up into spatial slices corresponding to events that happen at the same time. If we were moving at the same velocity, our paths would be parallel and this convention would yield the same carving into same-time slices. But if you are moving at some constant velocity relative to me, when you apply this convention, you will end up carving spacetime into spatial slices that are tilted with respect to mine. If I am team red and you are team blue, and the red (black) lines represent same-time slices at 1 second intervals ($t=1, t=2\ldots$) that I come up with and the blue (dark grey) lines represent the same time slices into 1 second intervals ($t'=1, t'=2\ldots$) that you come up with, our respective carvings projected into two dimensions would look like Figure 9.

What that means is that you will be regarding events that I regard as happening at the same time as actually happening at different times. And I will be regarding events that you regard as happening at the same time as actually happening at different times.

This fact will have lots of fallout and it is the source of many things that seem strange about special relativity. So, for example,

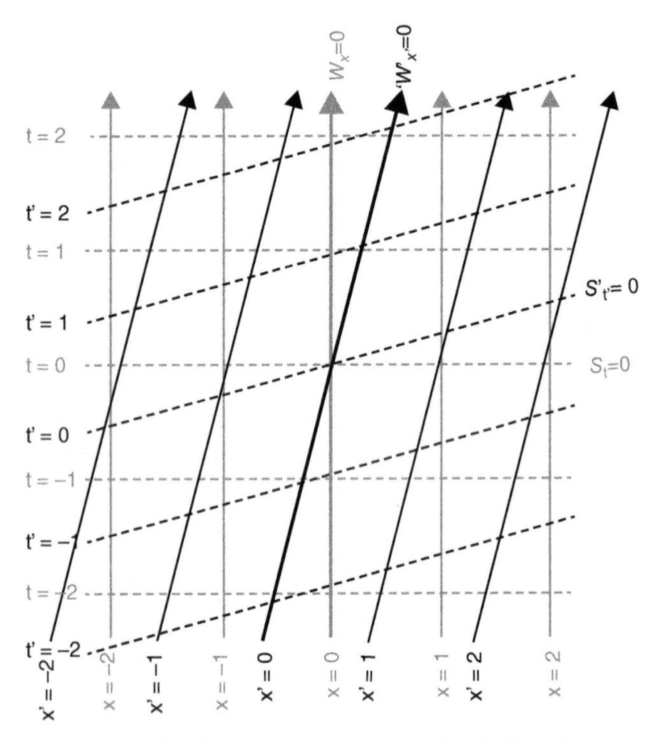

9. The lines in the diagram represent the coordinatization of spacetime from the perspective of observers moving at constant velocity relative to one another.

when I ask about the time it takes your clock to tick off an hour—that is, I will measure the temporal interval between the moment your clock ticks noon and the moment it registers 1 p.m.—it will be longer than one hour on my clock, and I will think that your clock is running slow. But the same will be true for you. If we reverse the procedure, use your carving into same-time slices, and let you measure the interval between the moment my clock passes noon and registers 1 p.m., it will be longer than one hour on *your* clock. It will also translate into disagreements about length. It is easiest to see how length is affected if you consider that measuring the length of an object is measuring the distance between its two ends at some particular time. If you and I are moving with respect to one another and we each measure the length of a metre stick that you are holding in your lap, for example, three things are true: (i) the stick is moving relative to me, (ii) what you are counting as the distance between its two ends at the same time is, by my lights, actually measuring the distance between its two ends at different times, and (iii) from my perspective, it has *moved* in the interval. To get its true length by my lights, I will have to correct for its motion in the time between the measurement of one end and the other, and the number that I get when I make the correction will be shorter than the one you assign. As above, the same will hold for you in the opposite direction when you are measuring lengths of the metre stick in my lap, and it will translate into different lengths for everything we measure with our respective sticks. These effects are known as time dilation and length contraction.

As bizarre as the idea may seem, it can be true both that your clocks are running slow by my lights and mine are running slow by yours. It is really at bottom just a result of confusing conventions for saying which events at different places are happening at the same time: that is, different conventions for carving spacetime into same-time slices. Although we are all measuring the same fixed spatiotemporal intervals between events, if we are moving at some fixed velocity relative to one another, we will divide those intervals into spatial and temporal intervals in different ways. Recall that Lorentz proposed explaining these effects by saying that there really were facts about spatial and temporal distances, and who is at rest and who is moving; he explained these effects by saying that when people are in motion, their clocks run slow and their measuring rods shrink. The genius of special relativity was to reveal that these differences are artefacts of conventions for carving spacetime into same-time slices. Once we get rid of those conventions, the intrinsic structure of spacetime emerges clearly, defined not by spatial distances and temporal durations, but by spatiotemporal distances. In philosophy, one often hears that physicists prize aesthetic beauty in their theories and sacrifice common sense to preserve the mathematical elegance. This is a paradigmatic example of the kind of mathematical elegance or naturalness that plays a role, and when you look behind to see the reasoning, it is much more compelling. As for common sense, which seems more commonsensical to you: the Lorentzian explanation or the Einsteinian one?

Any convention for carving spacetime into same-time slices is like looking at a striped wall through a water glass. If you and I are looking through different glasses, as illustrated in Figure 10, we see the same structure, differently distorted.

It is much cleaner, and less apt to confuse, to forgo these conventions and stick to describing the intrinsic geometry of spacetime speaking only about spatiotemporal distances, leaving aside any talk of spatial or temporal distances.

Since we are running through implications of special relativity that are strange to common sense, here is another one that is often discussed. Imagine a pair of twins that synchronize their clocks before they head out to space. They travel together until they reach outer space, then one stays on the space station while the other takes to a distant planet, travelling close to the speed of light, and returns after two years. When he returns to the space station he finds that although his own clock and body have registered the passing of only two years, his brother's watch has registered the passing of twenty years. And it is not just the watch. His brother's hair has grown twenty years' worth of inches; his heart has pumped twenty years' worth of blood; his eyes have seen twenty years' worth of troubles.

This implication of Einstein's theory was first described by Paul Langevin in 1911 and dubbed the Twin Paradox. It is different

10. Coordinate descriptions that impose a temporal ordering on all events distort the intrinsic structure of spacetime in a manner not disanalogous to looking at a striped wall through a glass of water.

from the puzzles above in that it is not an illusion stemming from using different conventions for carving spacetime into same-time slices. One sign of that is that it doesn't have the symmetry of those cases. Whereas above I say your watch runs slow and you say mine runs slow, in this case both twins agree that the travelling twin is younger. One would have to amplify the speeds to absurd levels to generate a difference of decades, but the theory does entail that by all inward and outward signs, there is a real difference in the age of the twins when they meet again. The reason for this is that what the watch on your wrist and the processes in your body actually measure is not the universal passage of time; there is no such thing. What they measure is the length of the path that you follow through spacetime. The travelling twin is younger because he travels a shorter path in the way that cars that take different routes between two points will measure different distances. (Don't be misled by trying to draw the paths of the two twins in space; it's a limitation of the diagrams that it will seem that the travelling twin has a longer path.)

A more realistic example of this effect is provided by fast-travelling cosmic ray muons created when highly energetic particles from deep space collide with atoms in the Earth's upper atmosphere. The initial collisions create pions which then decay into muons. The muons then travel down through the atmosphere and arrive at ground level with a frequency of about 1 cm^{-2} min^{-1}. Muons have a measured mean lifetime of about 2.2 microseconds. If the muons travel at nearly the speed of light then the distance travelled in a typical lifetime will be about 660 m. If the muons are produced in the upper atmosphere (15–20 km up) and travel, on average, a distance of about 660 m then not many should be capable of reaching the ground. The measured intensity of 1 cm^{-2} min^{-1} at ground level that we actually find is far too high. When we take relativistic effects into account, however, this is exactly what we would expect. Although the mean lifetime of the muon at rest is just microseconds, when it moves at near the speed of light, time dilation increases that by a factor of ten or more, giving the muons time to reach ground level.

Results like these are bizarre, judged by the standards of common sense, but so much the worse for common sense. They are direct consequences of the new geometry. The imaginative discomfort you may feel about them comes from the fact that old ideas about space and time are still organizing how you see the world. Common sense crystallizes regularities that hold across the range of everyday experience, but there is no reason to expect it to be particularly good at telling us what happens for people travelling long distances at speeds close to that of light. That's why we need physics, and if we follow where the physics leads, we have to be prepared to have even our most deep-seated commonsensical assumptions challenged. The strangeness of the world according to modern physics is just the feeling of a mind stretching its concepts to fit things outside the range of its own experience.

A bridge too far?

Among mind-bending possibilities raised by relativity, none has captured the popular imagination more than time travel. The idea of travelling through time has become such a staple of modern science fiction that it is hard to appreciate that it is a relatively recent invention. There have long been myths about people seeing the past or future, but the first widely known mention of time travel in a sense that involves travelling into different parts of time where you participate in events as they happen was in H. G. Wells's book *The Time Machine*. It is remarkable that Wells's book appeared in 1895, ten years *before* the paper in which Einstein introduced special relativity, because time travel seems to find such a natural imaginative home in the four-dimensional landscape of that theory. Wells's hero seems to be thinking in exactly these terms when he reasons that we ought to be able to visit other parts of time in the way we visit other parts of space.

It's easy to imagine from the inside what it would be like to travel back in time. A time traveller's own life would carry on as normal, but as she moves forward in the timeline of her personal history (with memories accumulating, hair growing, bones ageing, etc.), for at least a part of her history she would be moving backward in the timeline of the world. And it is easy to represent the trajectory that a time traveller would follow from the outside. In a spacetime format, it would look like Figure 11, moving up and then backward along the temporal dimension. Travelling into your own past and crossing paths with your younger self would look like Figure 12. Your trajectory would have you at the same point in spacetime at two different moments of your own personal history: the temporal equivalent of visiting the same place twice.

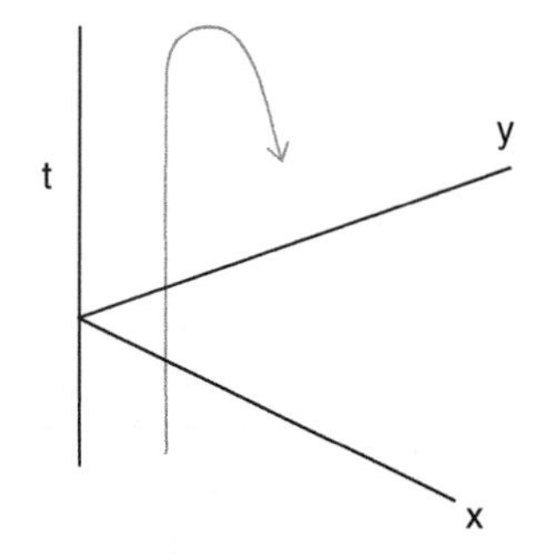

11. The trajectory of a time traveller follows a path that arches back on itself along the temporal dimension.

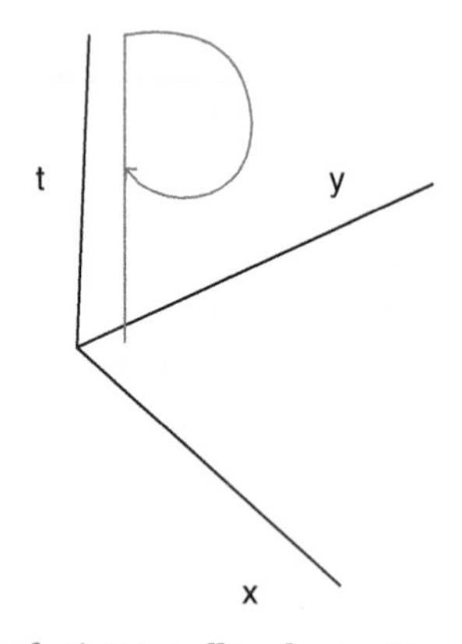

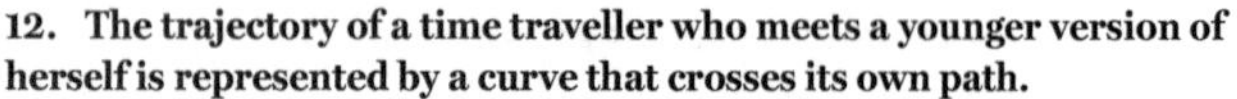

12. The trajectory of a time traveller who meets a younger version of herself is represented by a curve that crosses its own path.

Although we can draw these trajectories, there is a quite deep reason they can't happen in special relativity; and it is for a different reason than you might think. It is not because the past isn't 'there to be visited', so to speak. It's because you can't travel fast enough to get there. Travelling back in time would mean travelling faster than light. It is one of the deepest tenets of special relativity (although one that has come under increasing scrutiny for reasons connected to quantum mechanics) that nothing travels faster than light. The reason is that objects gain mass as they speed up and speeding up requires energy. The more mass, the more energy that is required. By the time an object reached the speed of light, its mass would be infinite and so would the amount of energy required to increase its speed. That is about as strong a ban on travelling to the past as you could find. Some have proposed that the ban could be circumvented by tachyons—particles that are created at speeds travelling faster than light so they don't need to be accelerated past that barrier. The existence of such particles, however, wouldn't get us any closer to the possibility of time travel for massive objects like us.

General relativity interestingly allows more leeway and has given rise to some fascinating real-science investigation into the possibility of time travel. Recall that according to general relativity, the geometry of the universe isn't generally linear and fixed, but rather curved and dependent on what sort of matter there is and how that matter is distributed in spacetime. Each solution to the fundamental equations of general relativity is (a mathematical description of) a way the world could be according to the theory. Time travel is possible according to general relativity if there are solutions to Einstein's equations that contain trajectories like those above that arch back in time. It turns out rather intriguingly that there are. For the reasons mentioned in Chapter 2, the equations are simple to write down, but very difficult to solve mathematically, so if someone asks whether such and such is possible according to general relativity, the answer is not always known. In this case, however, we know the answer because solutions have been found that allow travel into the past. The most famous of these is due to Kurt Gödel, one of the most famous and perversely creative mathematicians that ever lived. It is a particularly bizarre one and nobody thinks that it is an accurate representation of our own world, but it serves well enough to show that the general relativistic laws don't rule out time travel.

So does that mean that time travel is *really* possible? The fact that there are solutions to Einstein's equations that have trajectories that lead back in time hasn't convinced everybody. Many dismiss these solutions as physically meaningless because they think that when we try to describe in detail what those solutions are like, we end up with absurdity. Among those who thought so were Isaac Asimov and Stephen Hawking (along with many modern physicists), both of them citing an argument that has come to be known as the Grandfather Paradox. Here's Asimov:

> The dead giveaway that true time-travel is flatly impossible arises from the well-known 'paradoxes' it entails. The classic example is 'What if you go back into the past and kill your grandfather when he was still a little boy?'...So complex and hopeless are the paradoxes...that the easiest way out of the irrational chaos that results is to suppose that true time-travel is, and forever will be, impossible.

One version of the Grandfather Paradox is that someone in a fit of ancestral guilt about the source of her family wealth uses her money to buy a time machine and travel back in time to kill the grandfather responsible for the ill-gotten gains as a child in his crib. If he is killed as a child, he never grows up and performs the evil deeds that earned his money. But then he never has kids, the time traveller herself doesn't exist, and there is no money to buy the machine she uses to travel back in time. So it seems like the scenario gives rise to a contradiction. As interesting as the versions involving people are, they tend to confuse the issue by evoking intuitions about what we can and can't do that are drawn from everyday life. As a starting point, it is better to consider a version of the paradox that was proposed by John Earman, that sidelines those intuitions and involves nothing but mechanical instruments of a kind whose physics is more or less well understood. In this scenario, we imagine a rocket ship which can fire a probe which will travel into its own past. The rocket is equipped with a sensing device that controls a safety switch.

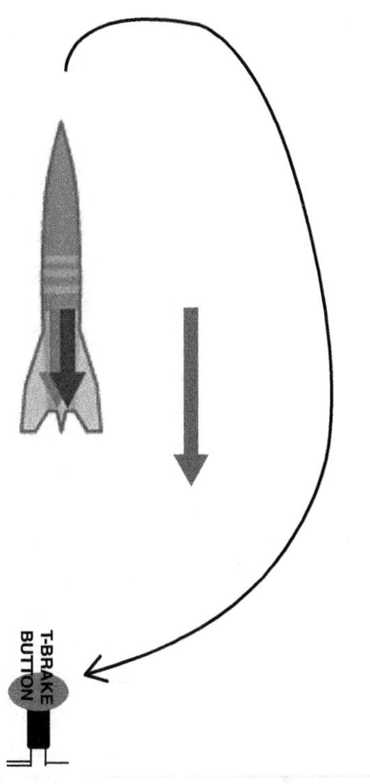

13. Earman's time-travelling rocket provides an example of a self-defeating causal chain analogous to the one that arises in the Grandfather Paradox. The rocket is designed so that when it is launched it will travel back in time and activates a safety switch that in its turn inhibits the rocket from launching.

The safety switch keeps the rocket from launching when it is activated, and it is turned on *if* and *only if* the 'return' of the probe is detected by the sensing device (Figure 13).

If the rocket fires the probe, the probe travels into the past, is detected by the sensing device, the safety switch is activated, and the rocket doesn't fire. If it doesn't fire, on the other hand, no probe is detected, the safety switch remains off and the rocket fires. So, the probe is fired if and only if it is *not* fired. It seems like either way, we have a contradiction, and if we have a contradiction, then time travel can't occur.

Notice, however, that we only have a contradiction if there is literally no logically coherent way that things could play out in this situation. So long as there is some way they could play out, there is no contradiction and Asimov and Hawking, together with anybody else who says that these kinds of situation lead to contradictions, are wrong. The physicists John Wheeler and Richard Feynman wrote a surprising paper in 1949 that pointed out that there *is* a logically coherent way things could play out in any situation that can be physically realized; indeed, an indefinite number of them. One way they could play out with the rocket scenario, for example, is that the probe launches but misfires, heading in the wrong direction away from its target, so it isn't detected by the sensing device. In another, it launches and heads in the right direction but the sensing device fails and the launch is not suppressed. In another, the probe doesn't launch at all, because something causes a safety switch malfunction and it remains on. You get the idea...in one way or another, some unintended glitch keeps the whole causal chain from completing successfully. Machine malfunctions happen all the time. Even if you set things up as carefully as you please, the physical laws will always leave a window open for the machinery to malfunction or an unanticipated contingency to come from outside and derail things. As long as it is physically possible for something to go wrong, we have all that we need to provide a logically coherent way that the situation can play out and the paradox is avoided.

The key here is that the kinds of systems that we are talking about are open systems, which means that they are embedded in a larger universe and can't be completely isolated from what happens outside. For any system like that, it is impossible to completely shield the system from external influences and so there is always the possibility of something not included in the description of the system interrupting the intended chain of events. There is nothing in our description of the rocket assembly, for example, that rules out a tornado, a lightning bolt, or an errant bit of fluff throwing a wringer into the whole operation. You could try to eliminate the possibility by including more and more of the world in your description of the system, but short of making the whole world into a single time travel loop, you can't get rid of the window for external influence that provides a something-goes-wrong-style resolution.

What about that scenario, then? Is there a way that things could consistently play out in a universe that consists of *nothing but* the rocket assembly described above, for example? Here things get a little technical. It turns out that in a class of cases that satisfy certain desiderata, we can find solutions that sit on the border between things malfunctioning and working fine and that turn out to be consistent. But considering worlds as wholes that form time travel loops moves us very far from the kind of setting that motivated our interest in the question. In the realistic setting which we are talking about, an open subsystem of a larger world—like a time-travelling bird or a rocket or a person—the Wheeler/Feynman observation is the relevant one.

This form of solution transfers to the Grandfather Paradox case in a straightforward way. The time traveller goes back to try to kill her grandfather and fails. That's what happens. End of story. There's nothing contradictory about that. There are lots of ways in which she can fail: maybe the gun catches, maybe her money lust overtakes her moral resolve, maybe she slips on a banana peel as she is about to pull the trigger. Grandpa grows up, begets son, who begets time traveller and the rest is—as they say—history.

We can switch this up in a way that mirrors the plot for time travel movies. This time you are the good guy. You go back in time to foil the plot that would, if it were successful, destroy the Earth before you were born. As surely as the remorseful time traveller fails to kill her grandpa, you succeed in saving the Earth. This is perfectly logically consistent and doesn't need to involve any violation of the physical laws. Good time travel stories know this and always involve surprising plot twists that serve to make sure everything comes out consistently. People can travel back in time to *try* to change history, even to change it in ways incompatible with their own existence, but something always happens to make sure everything comes out exactly as, well, exactly as it *did*.

Utilizing this loophole as a way of resolving the apparent contradiction is technically correct, but you wouldn't be alone if you think that there is something implausible and ad hoc about it. One way to try to bring out the implausibility is to point out that perhaps the time traveller slips on a banana peel on her first try at killing her grandfather but exploiting this loophole too consistently will involve a series of accidents that seem increasingly improbable. We know from the start that no time traveller can ever succeed in preventing her own birth and no self-defeating rocket assembly successfully completes a cycle. So we know that every time an attempt is made, something goes wrong. What if a thousand attempts are made, or a million, or ten zillion? Doesn't this string of motley failures, each individually improbable, together add up to something bizarre? We might not be able to show that time travel gives rise to contradictions but by pumping up the number of times we try and fail, we can show that it gives rise to strings of events so improbable as to be well-nigh impossible.

Again, here, things are not so obvious. Consider the self-defeating rocket assembly. While it is true that from one point of view the string of failures can seem anomalous, from another point of view it seems like just what you would expect. Looking *prospectively* from the launch, it can seem as though we see a string of unconnected and unexplained mishaps. Looking *retrospectively* from the point of view of the failure to inhibit the launch, we know some mishap occurred in each case and it is not surprising that we find mishaps when we look. It happens all the time in life that you try to do things you don't succeed at, and if you collect together all of the times in which you tried to do something and failed—if you select, that is to say, by the *outcome* of your attempts, cherry picking for the failures—you will have the same thing you see looking retrospectively from the launch. You fail every time, for unconnected reasons. Post-selecting by failure of outcome is unavoidable in a time travel setting because the starting point *is* the ending point. What is happening here is that the retrospective and prospective points of view are being brought into confrontation.

These two points of view are usually very different. You don't normally have information about the outcomes of your efforts in advance, but you do have that information afterwards and that is what is creating the sense of anomaly. That's a feature, not a bug. It's just the kind of abnormality that you would expect to happen on a circular trajectory in which every past event is *also* future. Sandra lives in Tucson. Normally she can go out without a winter coat. If you put her in Helsinki, she will need one. That's exactly the kind of 'abnormality' you would expect, given the difference between Tucson and Helsinki. And it's the same here. Time travel settings are different from settings in which there isn't time travel. In time travel settings, the past *is* future and the future *is* past, and no asymmetries between past and future can survive in those settings. We shouldn't be surprised that things that don't seem normal happen in a world in which there is time travel. We will be looking much more deeply in the next few chapters into why, and in what ways, our view of the future is different from our view of the past. That will make clearer why the prospective and retrospective seem to clash.

It is not hard to find other weird things that could happen if time travel were possible, things that would seem like magic to common sense. Consider, for example, a time traveller who steals a time machine from the local museum in order to make her trip.

Once her trip is complete, she donates the time machine to the same museum. Who invented the machine? She stole it from the museum, so she didn't create it. And the museum got it from her. Or imagine the writer who finds a manuscript in a box in his attic and publishes it, leading to riches that allow him to buy a time machine and travel back in time to plant a copy of the manuscript where it can be found. Who wrote the book? Again, not him, but it seems nobody else either. This all seems fantastical. Does it rule time travel out, or are these just abnormalities of the kind we should expect in worlds where there is time travel? Who's to say? Maybe it seems like magic to us *because* time travel doesn't happen in our world. If we can describe what the world would look like if there were time travel in a way that doesn't lead to contradiction or violation of physical law, then as weird as that world would seem to sensibilities formed in a world without time travel, perhaps it would just be normal to inhabitants of a time travel world.

Or maybe there is something that we haven't taken into account; maybe the laws of thermodynamics, for example, constrain the propagation of information in a way that would rule these things out. These are difficult, unsettled questions about which there is at present no consensus. If we separate things into the categories of (i) bizarre-seeming things that would happen in worlds in which there is time travel, and (ii) contradictions or violations of physical law that time travel would give rise to, we have found plenty of the former and none of the latter. There is no question that worlds in which time travel occurred would seem weird, but there's nothing that has been identified yet that seems to rule it out as a matter of logic or physical law.

That invites the question of whether our own world might have undiscovered pathways leading back in time, or whether we might be able to create such a pathway. Of course, we don't really know. People have examined whether the laws of general relativity might allow the manipulation of the geometry of spacetime in such a way as to create new paths that circle back in time. Current proposals involve dramatic alterations of the geometry of spacetime, so they are not the kinds of thing that we could feasibly engineer, but it has raised new questions and given rise to some fascinating physics. Of course, there are nay-sayers who say that we know that time travel will never happen in our futures, because if it did, we should expect there to be lots of travellers from the future around now. Who knows. We'll head back now to firmer ground and talk about things that are closer to the everyday world.

Chapter 4
The arrow of time

If you throw a tennis ball in the air, it will leave your hand travelling upward with a certain velocity, but gradually gravity will outweigh the upward force with which you threw it and it will reverse direction and fall back down to the Earth. The path it traces looks like Figure 14, going from the left side of the page to the right.

The mirror image of that trajectory in time reverses the sequence of positions through which the ball passes, starting on the right, travelling up and in the opposite direction, and landing on the left. We can obtain a similar 'reverse trajectory' for any physical process. Since any system is made of a collection of particles and the states of the system are defined by the positions and velocities of all of the particles of which it is made, reversing the order of positions through which the particles pass gives us the mirror image of the original trajectory in time. If the original trajectory takes us through a sequence of positions $S_1 \ldots S_n$, the reverse trajectory takes us through the sequence $S_n \ldots S_1$. (Since velocities are rates of change of position, reversing positions will also reverse the direction of velocities. Everything works out if you think of velocities as shadows of positions. If you change positions, the velocities will follow.)

Here's an interesting fact that may surprise you: the physical laws say that if the first trajectory is possible, so is the second. This is

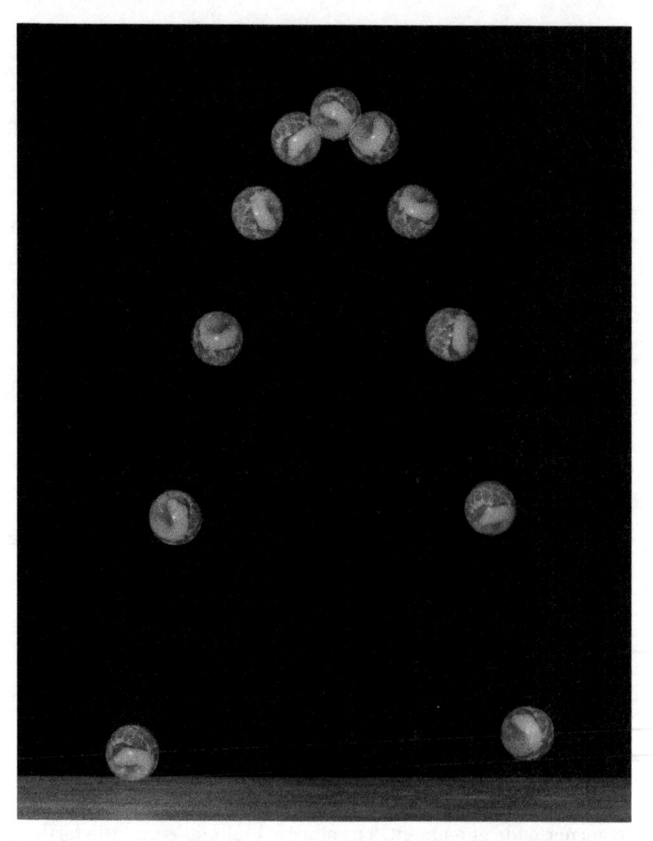

14. The trajectory of the bouncing ball is time reversible. If it can happen in one direction, leaving the left side of the page and landing on the right, it can also happen in the reverse, leaving from the right and landing on the left.

an explicitly derivable property of the Newtonian mechanical laws. It doesn't matter what kind of system you are talking about or what kind of process you have in mind; every system is ultimately made of particles obeying the same fundamental equations. If those equations say that the temporal reverse of every physical process is also physically possible, then anything that you see happening in one direction can happen in the other.

And *yet*... the everyday world is rife with processes that happen all the time in one direction and never in the other. Consider something as familiar as a dinner party. People sit at a table, laughing and gesturing while ice cubes melt in glasses of scotch. A duck egg cracked over steaming pasta is cooked by the rising heat. A guest stirs cream into a cup of coffee. A teacup slips from the edge of the table and shatters on the floor. None of the processes I just described—ice cubes melting, steam rising, eggs cooking, cream dispersing, a cup shattering—happen in the other direction. It might be true that the laws that govern the microscopic particles that make up all of these systems say that any process that can happen in one direction can happen in reverse, but anyone with a pair of eyes knows that's not true. This is a very deep puzzle that brings the mismatch between the fundamental laws of nature and the familiar world of everyday sense into sharp relief.

Research that began in the mid-19th century sought to resolve it. Unlike the leaps of the imagination that led to the theory of relativity, this story is the product of the accumulated insights of many contributors. There were many missteps along the way and it is only recently that it has come together in clear enough focus to garner wide agreement. The story actually starts a little farther back, when engineers figured out how to harness the untapped power of steam into the engines that drove the Industrial Revolution. Until the 1600s, horses and men provided most of the energy for industry. Scientists began tinkering with steam early to fire the furnaces of glassblowers, in 1698 the British inventor Thomas Savery patented a pump which he described as an 'engine to raise water by fire', and there was no looking back. The technology was improved over the next century and by the mid-18th century, steam engines were so powerful that 'horsepower' was coined to describe the number of horses an engine could replace. If your car has a 160-horsepower engine, you have the power of 160 horses under your hood.

With the development of steam engines, scientists began studying the relationship between heat and energy and the new field of thermodynamics (*thermo* = heat, *dynamics* = motion) was born. The first law of thermodynamics is a version of the familiar laws of conservation of energy. It states that although you can raise or lower the energy of a system by exchange with the environment, energy is neither created nor destroyed. The second law is more novel and it is the one that interests us here.

To understand the second law of thermodynamics, you need to know that there are two ways that gases can exchange energy with their surroundings: as heat and as work. Heat is what you think it is. It's the form of energy that cooks food, melts ice, and warms chilly feet. Work is energy transferred by a measurable mechanical effect. So, for example, a system does work on its surroundings if it pushes a piston or lifts a heavy object. Early versions of the second law were formulated in terms of heat and work. One, due to the German physicist Rudolf Clausius, stated that no process whose only thermodynamic consequence is the transfer of a given quantity of heat from a cooler body to a hotter one is possible. Another, due to the British physicist Lord Kelvin, said that no process whose sole thermodynamic result is to transform heat extracted from a source at the same temperature into work is possible.

Even though they sound very different, these versions turn out to be equivalent. They rule out exactly the same processes. And it turns out that, even though early research was focused on the properties of gases in enclosed containers, the laws of thermodynamics are fully general. They apply to all physical systems, whatever their composition. The modern formulation of the second law is a little more abstract and captures its sweeping generality. It also provides the basis for a much deeper

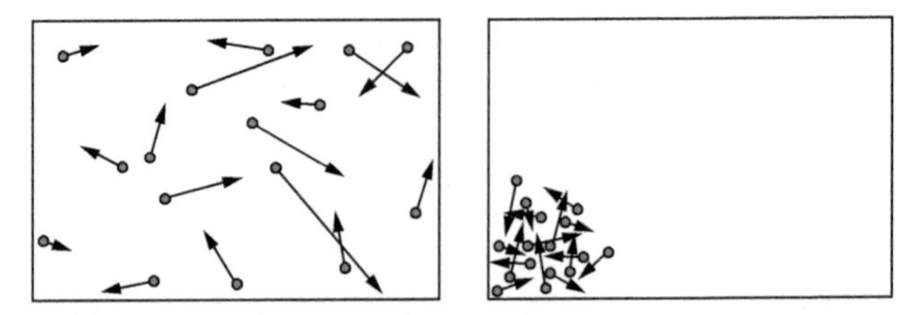

15. The particles of a gas in a high entropy state will be distributed evenly through the accessible space and moving in random directions as shown on the left. The right diagram shows a contrasting example of the particles of a gas in a low entropy state.

understanding of the phenomena of heat and work. It introduces a quantity called entropy that can be calculated for any system from its microstate. A system's microstate is a fully detailed specification of the positions and momenta of the particles that make it up; and you can think of entropy as a measure of the degree of disorder or randomness in the way its constituent particles are arranged and moving around. For every system there is a maximum entropy state. For an opaque gas in an enclosed container, for example, the maximum entropy state would be like the illustration on the left of Figure 15, with the particles distributed evenly and colliding randomly. To the naked eye, the gas would fill the container uniformly.

A low entropy state would be like the one in the right-hand illustration of Figure 15, with the particles all packed into one corner, or one in which the particles are arranged in stripes, or one in which they form a picture of the Italian flag. It is one in which the particles form a recognizable pattern or have a recognizable order. For cream in a cup of coffee, the maximum entropy state occurs when the cream is fully mixed in (Figure 16(a)). A lower entropy state would be one in which the cream was concentrated in identifiable swirls (Figure 16(b)).

High entropy

Low entropy

16. (a) A cup of coffee with cream in a high entropy state with the cream fully mixed in. (b) The same cup of coffee in a low entropy state. The cream in this case is only partially mixed and shows visible patterning.

The state of maximum entropy is called the equilibrium condition, because once a system enters that state it tends to stay there unless acted on. The second law of thermodynamics in its modern formulation states that the total entropy of an isolated system (one that is not interacting with its environment) never decreases. In other words, if any isolated system that is not already in a state of maximum entropy is left alone, it will evolve towards that state and stay there once it is reached. That should accord with your intuitions about how things generally behave: gas released into a closed container will tend to disperse to fill it evenly; cream poured into coffee will tend to spread until it is mixed uniformly; and in general whenever we have a system that has some macroscopic structure or order, that structure will tend to decay if it isn't actively maintained (at the expense of energy drawn from outside). The second law of thermodynamics turns out to subsume all of the known temporally irreversible physical processes from the cooking egg and melting ice cube to what are often described as the ravages of time: houses crumbling, paint fading, faces sagging, and bones decomposing.

With the second law of thermodynamics in hand, we have a sharp way of stating the puzzle that the time directedness of the everyday world presents to physics. We have one set of laws—the laws that were inherited from Newton and which govern the particles of which every physical system is made, and which are supposed to be complete and accurate—which state that any process that can happen in one direction in time can happen in reverse. And we have another law—the second law of thermodynamics—that patently contradicts that. It rules out the temporal reverse of all of the order-corroding, bone-decaying, heat-dissipating processes we see around us and it accurately describes the world as we find it. Something, clearly, is wrong.

The first hint of a clue about how to reconcile the blindness of the microscopic laws to temporal direction with the clear time directedness of the everyday world came from Ludwig Boltzmann, a German physicist with a Tolstoyan beard and a tragic personal story. Boltzmann began by distinguishing the microstate of a system, the complete description of the positions and velocities of its microparticles, from its macrostate, a coarse-grained description defined by the values of a small number of easily measurable parameters. These parameters average over microscopic properties of a system, specifying only the pressure, volume, and temperature of a gas. For every specification of pressure, volume, and temperature, there are a great many ways in which the individual particles of a gas could be arranged that would correspond to those values, and Boltzmann noticed that the division of microstates into macrostates is radically uneven. Some macrostates represent an enormous number of microstates while others cover only a few. The state of maximum entropy, it turns out, covers the overwhelming majority of microstates, while states of low entropy comprise a minuscule number. And in general, the lower the entropy of a macrostate, the fewer the number of microstates it encompasses.

Since we know the microscopic laws, we can determine how a collection of systems starting in any given collection of microstates will evolve macroscopically by just paying attention to changes in their coarse-grained state. If we start out with a collection of systems evenly distributed over the microstates compatible with a given macrostate M and let them evolve in accord with the microscopic laws, that will tell us how systems that start out in M tend to evolve from a macroscopic perspective. If we do that for the state of maximum entropy, pretty much all of the systems that start in that state stay there. I say 'pretty much', because there are some defectors. If you waited for a billion years, for example, you could expect to see a very small number of systems evolve momentarily to a lower entropy state and typically come back. If you blinked, however, you would miss them and over the course of a couple of millennia, you have almost no chance of seeing any defectors. So 'pretty much' means 'for all practical purposes, all'.

The microstate of a system in equilibrium is likely to be changing. The particles that make up a gas, for example, will be moving around and colliding; but not in ways that change its pressure, temperature, and volume, and hence not in ways that change its macrostate. If we take a set of systems evenly distributed among the microstates compatible with a macrostate of less than maximum entropy, and watch how they evolve, we find that pretty much all of them (in the same sense of 'pretty much') will be evolving towards states of higher entropy. Again there are some defectors, but they are so rare that we would have to wait much longer than our collective lifetimes before we could reasonably expect to see one. Boltzmann's insight was to appreciate that this isn't because the laws governing *micro*states have any direction built into them, but because the division into *macro*states is so radically uneven that even if systems are moving around randomly from one microstate to another and there's a mirror image of every microtrajectory, it is still going to work out, statistically, that the overwhelming majority of systems in a low entropy state are heading towards a higher one.

That is the truth in the second law of thermodynamics. It is not that it is strictly impossible for a system to go from a high entropy state to a lower one. It is just that it happens so rarely we would never expect to see it in real life. We need to add one more thing to complete the puzzle. In our world we see a great deal of macroscopic change, and we need to add an assumption that would explain why we see so much change. That assumption is that our universe started out in a low entropy state and has not yet reached equilibrium. We happen to live in an era of increasing entropy.

Picture a planet whose population is evenly distributed across its surface. Now imagine that it is divided into countries of different size: not just different like Finland/China different, but a difference that is millions of gazillions times that in size. Now let everybody wander more or less randomly so that they visit everywhere and spend the same amount of time in places of the same size, and so that for every person that travels a path from A to B there is another that travels the same path in reverse from B to A. If you think through what migration patterns would look like under those conditions it should seem evident that (i) at any given time, the overwhelming majority of people will already be in the biggest country and will stay there, and (ii) the overwhelming majority of people in any one of the smaller countries would be headed towards bigger ones. If the disparity in sizes was great enough to reflect the disparity in the number of microstates collected together by thermodynamic macrostates, the number of people heading from a big country to a smaller one would be close to non-existent. That is the second law of thermodynamics.

Now the final piece. If we *start out* with people distributed evenly over the face of the planet, we wouldn't expect to see much migration at all. The vast majority of people would already be in the largest country and would stay there. In our world, of course, there's rather a lot of macroscopic change, so we add an analogue of the assumption that our universe started out in a low entropy state and has not yet reached equilibrium. We crush everyone into one of the very tiniest countries so that they are packed in like sardines, and then open the floodgates, so to speak, and let them wander freely. Over time, they will spread out over the whole world filling it evenly. And while they are wandering to fill the space—in the period after the floodgates are opened before the world is filled—we would see a lot of movement across national boundaries pretty much always in the direction of larger countries. That movement across national boundaries is what we observe as macroscopic change.

The emergence of complexity and life

The understanding that the macroscopic asymmetries in our part of the universe may have their origin in a low entropy state in the distant past that develops under laws that don't recognize a difference between past and future is one of the triumphs of modern physics. It fits with everything that we know and embodies a quite wonderful set of physical insights. In physical terms, the right way to understand it is that microscopic processes are governed by laws that don't distinguish past and future, but if we carve the world up into the coarse-grained macroscopic categories discerned by our senses, and start the world in a low entropy state, those laws will entail that it will evolve towards higher entropy until it reaches the maximal entropy state. Since in qualitative terms, low entropy states are states characterized by a high degree of macroscopic order, to watch a transition from low to high entropy is to see a system lose structure, become disordered, and appear increasingly uniform.

The thermodynamic gradient—the name for this slope of transition from low to high entropy—is what creates conditions that form the backdrop for the emergence of the complex structures that populate the biosphere and eventually life. In the flux of transition from order to disorder, complex structures arise, at first spontaneously, and then by design. Parts are thrown together by accident and some of them bind together into stable configurations. Among the stable configurations, some prove able to do things that promote their survival and among those that survive, some find ways to make copies of themselves. Once we have systems that reproduce themselves, with some mutation, and a selection mechanism operates on these systems, the process of Darwinian evolution is launched. We don't have to draw a sharp line between life and non-life if we are content to start with something simple like a self-replicating molecule and see how Darwinian evolution over a sufficiently extended period will produce something complex enough that falls clearly on the side of life.

Evolution over long time scales is an effective search mechanism for complex systems that are able to survive and reproduce. This means a lot of different things. It means being able to meet their own energy requirements by converting energy from the environment into work. And it means competing for resources, resisting predators, and adapting to change. It turns out that a very valuable commodity in all of this is the ability to capture and use information. This is something that emerges from the evolutionary record rather than something one would guess just by looking at the fundamental equations of physics. The underlying physics makes possible all kinds of complex configurations. Selection pressures act on capabilities of these systems, that is, the things that they can do to keep themselves alive. The systems that get selected, survive, and are preserved by evolutionary dynamics—and hence the ones that populate our world—are the ones that utilize and process information in effective ways. It is sometimes said that the information age started in the late 1970s, but Nature discovered the utility of information long before we did. Major evolutionary transitions often involve changes in the way information is stored and transmitted.

Here's how this works with living creatures and how it relates to thermodynamics. Microscopically, everything that happens, happens in accord with unchanging microscopic laws that don't favour any direction of time. Because the universe has a low entropy past and hasn't reached maximum entropy, we see a lot of change in the direction of increasing entropy. The low entropy past makes it possible for information about the macroscopic past to accumulate in the form of records. The macroscopic world where there's a thermodynamic gradient is littered with records that contain the imprint of its macroscopic history. A footprint in the sand, for example, is the mark of a passing human. A half-melted ice cube in a glass of warm water is the remnant of a fully formed ice cube. A photograph, a series of letters on a piece of paper, all of the semi-ordered systems you see around you are evolving from states of even lower entropy, or greater order, and bear the imprint of their past. The macroscopic development of those systems, left to their own devices, will evolve towards states of greater uniformity: footprints will wash away, ice will melt. But evolution populates the universe with open systems using energy from the environment to maintain their own integrity, dissipating heat, and using information to guide behaviour.

The information contained in the macroscopic environment (i.e. in the present state of systems evolving from lower to higher entropy) is available to *other* systems to use as a basis for their own behaviour. If it is not immediately obvious why the ability to read information about the past is a selective advantage, think of it this way. A deer that can read the signs of a recent lion hunt, or a predator who knows that a beaver dam means a beaver, will do better than one that doesn't. A creature that knows what happened knows (something about) what is going to happen. I've worded this in a casual way above. Being more careful, one would not say that it is the creature that reads the information; one would say that it is us understanding that the reason that nature selected *that* response to *this* trace (or the reason that creatures who produce that response to this trace survived) is that this trace carries information about something to which that response is appropriate. And that is the clue. Because there are macroscopic regularities that link what *has happened* to what *will happen*, there is a selective advantage to utilizing the information in the environment. Like the deer and the beaver, we rely uncritically on records for information about the past in our daily lives. What we learned about thermodynamics and the low entropy past makes the physical facts that underwrite that reliance explicit. It is because the universe started in a low entropy state that our macroscopic environment has so much order, and why that order bears the imprint of an even more ordered past.

Agency

If we step back, and take a cross-section of our world, ordered by scale, and look at the layers of structure from the microscopic all the way up to the level of the human being interacting with a macroscopic environment, it would look like this. On the bottom, there would be the geometry of spacetime, which imposes (or embodies) constraints on the causal connectability of events. Then there are the microscopic laws that are blind to the difference between past and future and tell us how the matter of which all things are made behaves. In the presence of a low entropy past, those laws give rise to the macroscopic asymmetries embodied in the second law of thermodynamics. The macroscopic asymmetries in their turn support the emergence of creatures that use information to guide behaviour. Over time, these creatures evolve following a developmental trajectory that leads from simple forms that respond to local traces in the environment (the smell of prey, the sound of danger) to ones with increasingly sophisticated ways of tracking and processing information. The existence of longer-term and more abstract regularities gives an advantage to creatures with memories that can track those regularities and exploit the information they contain, and so those are the creatures that survive. We are the product of that developmental trajectory. We are soft and slow and weaker than we should be for our size. The factor that allows us to survive is that we have some very special equipment for gathering and utilizing information.

All living systems, to the extent that they use information to guide behaviour, feed off the thermodynamic gradient. Some systems rely on the low-lying fruit, wiring behaviour to information-carrying triggers in the environment (gazelles respond to smells left behind by passing predators, ants will follow pheromone trails leading to food). Human beings go after the sweeter fruit in higher branches. We accumulate large bodies of information, process it in complex ways, and put the information to unlimited, flexible use. In a human being, it is not the information gathered over evolutionary time and built into the hard structure of the body, but the information gathered in personal time and embodied in the soft structure in the brain, that guides behaviour. This opens into the topic of Chapter 5, but there is one last observation that I want to make before moving on.

If we allow the idea that important transitions happen when new ways of capturing and using information are found, it is easy to see the emergence of culture as entirely continuous with this development. In a world like ours the collective environment is littered not only with natural artefacts like fossils and footprints, but culturally accumulated records like books, libraries, and scientific databases. These are the amassed products of a history of creative exploration placed (literally nowadays) at our fingertips. They contain not just information about the past, but records of what we learned from it: *lessons* and *inventions* and all of the creative bounty of a collective history that we have had the wisdom to preserve and pass on to our children. It includes what we have learned about agriculture and aeroplanes, medicine and music, philosophy and politics. It includes not only the intelligence embodied in books, but the methods we have developed to exploit it, and the practices and institutions that make its transmission possible. A moment's reflection will make it clear that the lion's share of our power as a species comes from our access to this accumulated storehouse of knowledge. Wipe it away, and we are delicate and vulnerable.

Again here we see the same pattern: new forms of complex organization arise when information previously allowed to disappear, to fade like footprints in the sand, is captured and used to do causal work. The process continues so long as information provides a selective advantage. The environments that reward intelligence contain enough change to make simple programmed behaviour ineffective, and enough regularity to reward the collection and analysis of large bodies of information. Culture gives us a much longer horizon, allowing us to accumulate information, capturing regularities that span generations, and developing more powerful techniques for processing it. It makes possible the development of social structures that allow science and technological enhancement to flourish. It also allows the accumulation of artefacts like compasses and calculators: smart devices that lighten our cognitive load. This is now looking more recognizably like time as we know it: the time of human history that we began chronicling with the universal dating systems described in Chapter 1, which forms the backdrop against which we live out our own lives.

Chapter 5
The time of human experience

Many people, when they encounter the relativistic image of time, are puzzled by the fact that it looks like a static block of events. Many in the physics community, including Einstein himself, thought that the new theory vindicated the Parmenidean view that the passage of time is an illusion. Others reacted with a sharp rejection of the new conception of time. A fierce debate between the French philosopher Henri Bergson and Einstein broke out in 1922. Bergson argued vigorously that the relativistic conception of time left out everything essential to time as we know it. After all, where in this image do we find the flux and the flow, the continuous change? Where do we find the openness of the future, the fixity of the past, and the unremitting passage of time?

This has been the central question in the philosophy of time ever since, and the whole issue has been engulfed in controversy. Part of the difficulty in addressing it stems from the fact that it is very hard to articulate in a non-metaphorical way what it means to say that time passes or that it flows. But it is also hard to convince yourself that there isn't something absolutely fundamental to our experience of the world that those metaphors are evocative of. Our pre-scientific beliefs about what there is on the surface of Mars or what the universe was like 2 billion years ago don't have much authority because these are things well outside the range of our experience. But time is something we know as intimately as we know ourselves. What physics tells us about time, if it is correct, has to answer ultimately to our own experience. And if physics is not going to be just a tool for predicting and controlling nature—if it is, in particular, to provide us with a way of understanding ourselves and our place in the cosmos—we could hope for an illuminating account of why, if the time of physics is so simple and austere, the time of human life is so intricate and tangled, so layered and complex.

Not many years ago, the gap between the familiar time of everyday sense and time as it appears in the relativistic image seemed unbridgeable. I think that is no longer true. The key is to resist the temptation to think that you understand what relativity is telling us about time by adopting an imaginative point of view looking down at the Universe from outside. That perspective is strongly encouraged by the visual images we use to convey the content of the theory. Figure 8 showed us the cosmologist's representations of the Standard Model, and if you look through any textbook on relativity, you will see spacetime diagrams everywhere. Recall how the diagrams work: they suppress one or two spatial dimensions to create a low-dimensional representation *of* (space)time *in* (space)time. We are very good at grasping patterns that we can see. By rendering spacetime in a concrete visible form these diagrams let us use our spatial intuitions to understand what the theory is saying about the more abstract four-dimensional structure.

But they are misleading in a way that is philosophically pernicious. One looks down at the diagrams and feels inclined to say, 'It's just a static block of events, there is no change or flow or flux.' When we do that, we are judging from the point of view of the time in which the image is embedded, not from the point of view of the time that is depicted in the image. That is a mistake. It is like looking at a musical score of a Bach cello sonata, for example, and saying, 'It is just a static block of notes; there is no change or movement.' While it is true that the score is a static thing—a structured set of notes drawn on a page—there is a complex sequencing internal to the music it represents. There is movement and flow and change *in* the music, and that movement and flow and change is captured in the score. If we want to understand what relativity is telling us about the time of lived human experience, we should be judging from the point of view of the time *internal* to the image as it is seen through the eyes of a person whose life is represented by a line, or a sequence of events laid along the temporal dimension.

How would we expect the four-dimensional spacetime of a relativistic universe to look to such a person over the course of her life? It will be helpful here to say a little bit about the human mind. Schematically, you can think of a human mind as a kind of information-processing system. It creates a special sequestered environment where information coming in through sensory channels is not used immediately to guide behaviour and then discarded (as it might be for an especially simple kind of animal that reacts to informational triggers in the environment), but is instead captured, stored, and used to modify a continuously evolving view of the world in its full spatial and temporal extent. When we ask how the world appears through the eyes of the temporally immersed human being at a particular moment, we are asking what the world looks like from their point of view at that moment. What do they know about other places and times? What are the thoughts, emotions, beliefs, attitudes they have at that moment? When we ask how the world looks through the eyes of the temporally immersed human being over some stretch of time, we are asking how these things shift from one moment to the next over that stretch of time. How are they transformed as the person moves through the world perceiving, thinking, acting, and feeling?

If we want to illuminate our experience, attention is focused on what bubbles up into the stream of consciousness, but we can't talk about that without also understanding something about what is going on in your brain below the level of awareness. A wealth of the most fascinating and complex science since the mid-1950s has given us a peek under the hood at the processes that give rise to and support conscious experience, and that research has been particularly instructive for understanding the perception of time. Pre-scientific common sense tends to assume that perception is like a movie camera and that it conveys an unfiltered, real-time view of the world.

The real story is very different. Between the time that signals coming from the environment hit your retina and the time you have a visual experience, your brain has done a lot of work. It has sifted and sorted, got rid of noise, and interpolated rather broadly. It has combined the visual signals with information coming in through other sensory pathways and used what it knows about how things generally hang together to organize it all into what presents itself to you as immediate awareness of an objective world.

The idea of this world has built into it a spatial ordering, the presumption that you can touch what you see, expectations about how things that present a certain way visually should feel if touched. Run your fingers over the stem of a wineglass and ask yourself whether there's anything intrinsically that connects that pattern of light and colour to this pattern of tactual experiences. You take those connections for granted because your brain works them out for you and they get built into the concepts with which your mind interprets your experience. You walk along a mountain trail looking around, and your brain uses visual cues to guide the movements of eyes, head, arms, and legs. It works out that the darkness in the upper corner of the visual field means that there's a tree with low branches and that the shadow in the lower left means there's a dip in the trail. It works out what signals it needs to send to the motor pathways to keep your head out of the way and legs adjusting to the shifting landscape. You aren't aware of what it takes to work any of that out. It seems as though *what you see* is unfiltered awareness of the naked world. In fact, what you see is an interpreted image that embodies what you need to know for the purposes of regulating action. Similarly, when you hear someone speak in your native language, you don't hear an uninterpreted stream of sounds; you hear *what they are telling you*. Your brain decodes the sounds without any conscious inference on your part, so that they come to you as *laden* with meaning.

Some surprising research connects all of this specifically with the perception of time. It turns out that one of the things that happens while the brain is sifting and sorting in the process of decoding sensory signals is that information about what happens over a short interval of time is carried forward so that what you are seeing at a given instant is not a snapshot of where things stand at that instant, but a summary of how things have changed over a fraction of a second. There's an easy way to convince yourself of this. Find an old analogue clock with a second hand and stare at it for a couple of minutes. In that time, the minute hand and the second hand will both have moved, but you know that in very different ways. You can tell that the minute hand has moved by comparing its position at the beginning and end of the two minutes. In the case of the second hand, you don't need to compare and compute. You *see* the movement directly. It has a direction and a speed. The movement itself is *directly visible*. Since motion is change in position over time, what that means is that what you are seeing at a given instant cannot be an instantaneous position, but something that spans a short interval of time.

It sounds odd, but there's nothing particularly mysterious about this once you've understood that there is a lot going on in your brain to interpret the sensory signals before you are presented with a visual experience. Your brain is simply integrating experience over an interval and incorporating that information into what you see, so that at any given instant you are not just seeing the state of things at that instant, but a little bit of the past. There is also evidence, interestingly enough, that the perceptual system uses what it knows to project a trajectory into the future, so that what you see is not just what is happening at an instant, nor even a bit of the past, but also a little bit of the (expected) future.

This makes perfect sense when you think about the fact that the point of perception is to guide action. Perception needs to get you ready to meet an expected event. The outfielder running for a flyball needs not just to see where the ball is but to have a sense of where it is going, and his brain is doing that calculation for him. In general, the more of the processing that can be automated in a quick and efficient way by the brain below the threshold of awareness the better. The conscious parts of your mental life are reserved for the slow, laborious inferences that draw on unlimited bodies of information and demand creative input. They are for making long-term plans and executive decisions, not managing the day-to-day activities that keep you alive. One of the central insights that emerged from cognitive psychology is that your mind—conceived as an information-processing unit that mediates sensory signals and motor responses—does rather a lot that you are not directly aware of, and we understand ourselves much better if we appreciate that what we experience is the visible tip of a much larger iceberg. Perception is not unfiltered awareness, but something that has undergone processing that builds in whatever information might be useful for regulating action. In this case, it builds in the speed and direction of motion over a short interval.

That is what explains the feeling of flow or flux that pervades your experience, that is, the sense that the world is always ongoing or in process, moving from past to future, perpetually in transition. It captures part of what the people on Bergson's side of the debate with Einstein felt so strongly was missing from the relativistic image of time. That missing piece has been given many evocative names over the years: it has been called the Moving Present, the

Travelling Now, the Whoosh of Experience, the Surge of Process, or as the philosopher C. D. Broad memorably put it (in a passage meant to mock its defenders): 'the flow and go of very existence, nearer to us than breathing, closer than hands and feet'.

This is an important step in understanding the character of our experience, but it is only part of the story. Perception doesn't occur in a vacuum. It feeds into a psychological context lined with memories. Memories come in many forms. There are images and sounds, episodes and stories. What we select and how we store things in memory is a very personal thing. My own mind is littered with fragments of conversation, glimpses across rooms, long rolling episodes recalled wholesale alongside snatches and titbits salvaged from the passing tide of time. These are like so many snapshots and old letters, wine labels and restaurant receipts, stuffed in a diary and arranged around the timeline of my life. The contents of memory grow over time, as we have new experiences, and they are always being reorganized and retrospectively reconsidered. New configurations of events emerge into prominence, others are pushed into the background; new experiences alter the significance of old ones.

Our memories give us glimpses not just of the things that happen to us, but of past selves. They are like slices of life. They give us a record of how everything seemed to us at different points in our lives. When I think about the first day at a summer basketball camp before high school, I remember not just the look of the air and the salty smell of sweat; I remember not just the feeling of skin on the pink edge of burn; I remember the feeling of a young body at full strength. I remember what it was like to be a country girl from rural Canada without any special athletic ability surrounded by urban American kids who seemed to have been born with a basketball in their hands. I remember embarrassment, uncertainty, and the elation of realizing the world was much bigger and more exciting than I had imagined. I remember, that is to say, not just what it was to be *there*, but what it was like to be *me*. If the sense that time *flows* has to do with the way things feel at a moment, the sense that time *passes* has more to do with a comparison between what things seemed like from different temporal perspectives in your life. You can take slices at different times and compare how things looked along those slices. You can compare in your mind's eye how things looked while you were 10 to how they looked at 20 to how they look at 40. You can look back at your own life and retrace the years in your mind, from early childhood, through youth, to today. When you do that, the years rush by like houses seen from a moving car. Such is the passage of time.

Not every kind of animal has the kind of memory that gives it awareness of the passing of time, because not every kind of animal remembers its past in the same way that we do. Not every creature keeps a running record of its life, is always looking backwards and forwards, and remembers what it was like to be itself at different times. It used to be commonplace to say that this kind of memory (called autobiographical memory) is uniquely human, but not nearly enough is known about the psychological lives of other animals to say so with any confidence. It is hard not to see the mother penguin cuddling her chick after a long separation—locating him implausibly in a sea of thousands of chicks—as not involving some kind of memory. Studies have shown the adults and chicks recognize each other's calls among the noise. But what is going on in her mind as she brings him in close (Figure 17)?

Is she just following an acoustic call she is programmed to recognize in the way that we eat when we are hungry, or is she thinking, 'This is the child I carried in my belly and whose future I imagined while I felt the early stirrings of its young life?' These are two ends of a spectrum with a vast range of possibilities in between. We don't understand very well what an animal's inner life is like until we know how it relates to time; whether it stores information about its past and in what form, whether it has representations of the future, whether it has plans and projects

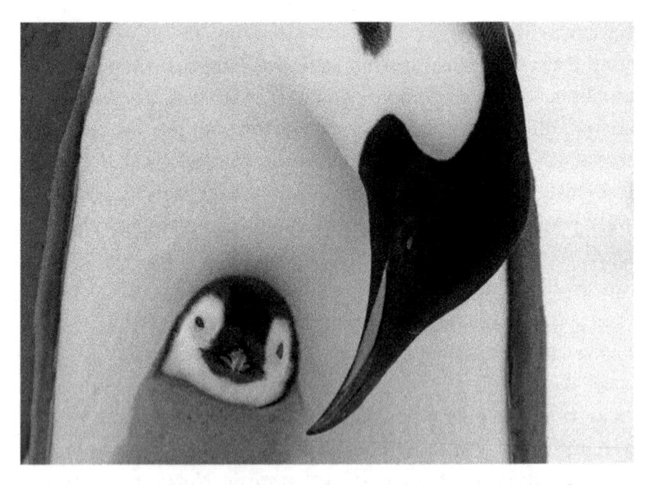

17. It is impossible to understand the inner life of other animals without knowing something about their relationship to time. Whether and in what form the mother penguin in this image represents time will have a great impact on the nature of her attachment to her chick.

like our own, or proto-plans and proto-projects of some other kind. These are things that are largely unknown for most animals. Research on octopuses, bees, scrub jays, and dogs tells us that they have rich cognitive capacities, and researchers are beginning to probe questions about what kinds of temporal representations they have, but there is such a wide range of possibilities, these things are difficult to study empirically, and there is little generality. Each animal has its own kind of mind.

One of the reasons it would be fascinating to know what kind of memory the penguin has is that it would tell us how much a penguin's experience is like our own. Our own experience is so pervaded by memory that it would be almost unrecognizable without it. One way to call attention to how memory structures even a brief episode of human experience is to think about what it is like to listen to a piece of music. Again, this is an experiment you can do yourself. Go and sit alone in a dark room with a pair of headphones. Bracket out everything else so that you are aware of nothing but a simple stream of sound. Put on something you haven't heard before—something simple: a single instrument or just a voice—and listen. There is silence for a few seconds, and then a single note. As that note sinks in, you think, 'ah... it's a cello' or maybe 'ah, a female voice'. That gives rise to some expectations about what will come next. As the note fades, a second note is registered and added to memory. Your earlier expectation is contradicted or confirmed. A new note is registered, compared against the expectation from the previous cycle, added to memory, new expectation is generated, and new, more definite expectations begin to take shape. The cycle repeats, with memories accumulating, and expectations becoming more definite at every stage. The mind begins to discern patterns, recognize motifs. It jumps ahead and completes a theme before the notes register. It is either satisfied or surprised by what it hears, delighted or disappointed.

At the first stage, the mind registers a note and forms a very indefinite sort of expectation. There's nothing at this stage yet in memory. At the next stage, the note and expectation registered at the first stage are incorporated into memory and form the psychological backdrop against which the second note is heard. A newer, more definite expectation is formed that draws both on the note that is being currently registered and the contents of memory. And so it goes, at each stage, the contents of the previous stage being incorporated into memory, a new note being registered, and a new expectation formed that draws on the whole accumulating stock of information being registered perceptually and incorporated into memory.

The sort of system that keeps an evolving record of its past and forms expectations for the future encounters every note as a partial revelation of an extended structure that will be eventually apprehended in its entirety. That makes a difference to the quality of the experience. The mind that confronts a theme for the third or fourth time hears it differently from a mind that confronts it for the first. Surprise, recognition, disappointment...these emotions all have a feel of their own. By the time you get to the last note, you approach it with the full memory of the preceding notes, and not just of the preceding notes, but the memory of the expectations confirmed, the surprising turns, the pleasing recurrences. In a satisfying piece, you experience the last note as a resolution of a whole unfolding structure. There are no loose ends, no promises unfulfilled.

A human life has that same kind of temporal structure, but with immeasurably more complexity, and it is complicated by the fact that we aren't just spectators to our lives. How our lives go depends in part on our choices. A life starts out, like a piece of music, with no memories and what presents itself as a wide-open future of indefinite possibility. As life goes on, you accumulate memories; the possibilities for your future begin to look more definite; you make choices that close off old possibilities and open up new ones. In the middle of your life, you have a large body of definite memories and a fairly clear vision of your future: memories behind you and as many ahead of you waiting to be made. Later years are heavy with memory and less burdened with future.

We can explore the view from a particular moment, or the progression of views over a life as a whole. The structure that is present at any moment has a temporal depth that comes from the fact that you've passed through all of the perspectives leading up to it. Your sense of who you are stretches backward and forward in time. Even if you feel little connection with the child or youth you once were, she is part of you. You carry memories of her experiences and expectations from the inside. You have a window on her inner life in the way that nobody else does, and her choices made you who you are. Some of us more than others, but all of us in our own way, carry our pasts with us. This sense of temporal extension carries over to others. Your sense of who other people are also tends to stretch the span of the time you've known them and to have the structure of perspectives drawn from different times layered over one another that your own experience does.

Look in the eyes of a father at a child's college graduation and you see the years of layered perspectives through which he is looking at his son. It is not just pride in his eyes, it's a mix of nostalgia and hope, maybe relief. You see the layered perspectives of the six-pound infant in his arms, the triumphant toddler who read his first word, the sweet third grader struggling more than he should have to with mathematics, the defiant teen.... Or look in the eyes of the daughter as she rages silently at the condescension of a doctor treating her ageing parent. She sees what the doctor sees: the tiny woman, all of her attention bent on a struggle with the button on a loose sweater. But she also sees what he doesn't. She sees that woman as a young wife in a foreign country with two kids who belong more quickly and more fully than she does. She sees that woman in her later years at full strength in command of a classroom, sees her coolly coping for decades with a disease she didn't deserve, and over the years as the quiet spine of an unruly family. Her view of her mother isn't clouded by her memories; it is enhanced. It is more expansive. She sees the full length of her mother. She sees the whole of her.

The events of one's own life are encountered from multiple perspectives, first in anticipation, later *in praesentia*, and finally in retrospect. And it is not just the events, but the perspectives themselves that we represent. Later perspectives have earlier perspectives as constituents. Earlier ones have later ones as constituents. (Many people anticipate remembering occasions like weddings or graduations so earnestly that they design the whole event in order to make good memories.) We build up a progression of views over the course of a life by stringing together these momentary perspectives in the order in which they occur: first this, then that, now this... And then we look at how the whole body of memories, perceptions, hopes, and fears changes over time. Looking forward from any point in your life, you see an open future of branching possibilities. Over time as you make choices and experience the effects of your actions, you see these transformed into the thin hard line of fact. Our inner life—with all of the evolving expectations and emotions that attend our experience—is governed by this dynamic.

There is nothing illusory about the experience of time passing, nor of the flux, the flow, the surge of life. The continuous cycle of anticipating, then experiencing, then remembering events is perfectly real. We don't look at time from the outside. We experience it from the inside, with all of the tension and excitement that comes from seeing it unfold in real time. It happens one thought, one idea, one action at a time, under the guidance of our own choices. It has its surprises and disappointments, its aha moments, its false starts, and sudden flashes of insight. And it has an internal order defined by the way that memories of the past inform the present, and present decisions inform the future.

Let's return now to the question of whether we can reconcile our own experience with the relativistic image of time. We live all of this as it happens, one moment at a time, but if we lay it all down in a diagram with one of the dimensions representing time so that we could see the parts and their relations to one another at once, as we do when we write down a timeline of a story or the notes in a piece of music, we see all of this structure—the memories, the expectations, the experiences—rendered by the way that our perspective changes from one slice of the line to the next. This gives us an internal sense of time defined for the psychological history of a single observer.

Communication among observers (and the creation of time-keeping technologies like clocks and watches) serves to stabilize an intersubjective notion of 'What time it is' well enough for practical purposes. You talk to other people, sending information back and forth so that their words and actions are woven into your life and you experience their lives unfolding in tandem with yours. This gives communicating observers, moving slowly relative to one another, a shared sense of the concrete reality of their present, the fixity of their common past, and a sense of charting a shared path into the future. And this whole unfolding of histories—with your life woven into it, and the knowledge that your life will be history to those coming after you—that is, that they will live in a world that bears the traces of your actions—is the Universe according to relativity theory. Relativity doesn't present us with a changing image of the world, but it seems to me as real and satisfying as an image *of* change as one could want. And the aspects of our experience of time that seem to be disorientatingly absent in the relativistic image—the flow, the flux, the passage and change—those are captured, just in a more subtle way.

When you've lived long enough to see events from multiple perspectives, there are moments when time can seem to break down. These are moments in which you pause and, to use Virginia Woolf's phrase, time seems to 'taper to a point'. In those moments, the richly layered temporal perspectives distributed across your life come together. You seem to live them all at once: triumphs, sorrows, humiliations, the successes and the failures. You feel you can touch the distant yesterdays and that they were already then pregnant with the present. Temporal perspectives are laid out along your life, and you pass through them one by one, but they are also—in the form of memories and anticipations, and memories of anticipations, and anticipations of memories—present like shadows in every part of it.

Chapter 6
The big picture and new horizons

It is time now to pull the pieces together. Questions about the nature of time have been at the heart of philosophical thinking for as long as we've had a record of its history. These questions passed into the hands of the scientists when the structure of space and time became connected to motion and made part of the subject matter of physics. The history of physics, from Newton through to Einstein's two revolutions, wrought changes in our conception of time that we couldn't have anticipated from the armchair. There are parts of physics that are in a state of confusion, but this strand of development is a story of philosophical illumination and conceptual beauty. The discussion here provides an opportunity to see what distinguishes the methods of physics from those of philosophy. Those methods are evident throughout this book. They prize data and precision. They are ruthless with common sense and carve away structure (such as the difference between being at rest and travelling at a constant velocity, or the distinction between being at rest in a gravitational field and accelerating upwards in outer space) that is not manifested somehow in the observable motion of material things.

Here, in broad strokes, is what physicists have taught us about the bare structure of space and time and what we need to add to that structure in order for time to emerge in a form that is recognizable to us as the familiar time of everyday life. Geometrically, time is one dimension of a four-dimensional structure that includes space. Although the microscopic laws that govern what happens at small scales say that every process that can happen on one temporal direction can also happen in the other, they also entail that any system in a *macro*state of low entropy will (with overwhelming probability) be evolving towards one of higher entropy. The time directedness of the everyday world is explained by the fact that our universe appears to have started in a very low entropy state and has been evolving towards a state of higher entropy ever since. Along that path or slope from low entropy to high entropy, information about the past leaves traces in the present: concrete records of past events that are locally available in the world. On our planet, there happened to be the right chemical ingredients and conditions that provided an opportunity for them to combine in a way that allowed the emergence of life. An evolutionary process that favours the exploitation of information gave rise to creatures that survive by using information to guide behaviour. With such creatures on the scene, we have not only change and growth, but *witnesses* to change and growth, and eventually the kinds of witnesses that can pull the whole story together: remembering the past, anticipating the future, and trying to steer the change locally as it is happening. Those creatures are, of course, us.

Relativity taught us that some of our naive ideas about time are wrong; in particular the idea that the universe is a large, spatially extended substance that evolves in time. Instead, we learn that the universe is a network of events. Chains of these events correspond to the histories of objects. There isn't a well-defined temporal order for the network as a whole, but we can always choose some system *in* the world—a rock, a lobster, a person—and follow how things unfold from its perspective. There will be asynchronies between how things unfold from the perspective of systems at a spatial distance from one another. Like the ships on the open water who can only coordinate their timelines by sending signals back and forth, they won't be able to establish relations of simultaneity across their timelines. Strictly speaking there won't be a fact about exactly what was happening to my friend in Milan at the exact moment I snap my fingers in New York, but those asynchronies aren't noticeable at the speeds and scales relevant to our own lived experience. They become important at astronomical scales and speeds close to that of light.

It's helpful to see how all of this looks at different scales and from different perspectives. If we are primarily concerned with what all of this says about us and how we should think about our own lives, we have to scope in closely on an individual human life and look from the inside out. From this perspective, we see our lives unfolding in a temporally asymmetric way from a remembered past into an open future. If we are interested in our universe from a cosmological perspective—we scope way out and model spacetime as a whole. From such a perspective, there is no unfolding. The Universe *includes* time; it is not an object that unfolds *in* it.

Scoping in closely on the microscopic level, one era is just like the next. It's all just particles drawn from the same unchanging stock, obeying the same microscopic laws. If we scope *way* out to a sufficiently coarse-grained perspective (one that pixilates the universe into cells 54 million light years across), matter is distributed roughly evenly across the universe and it looks the same in every direction. If we focus on our planet and zoom in again, not quite to the microscopic level, but to the macroscopic level to reveal the world of our own lived experience, we see something quite different. We see a world with a complex variegated structure, populated by things like forests and rivers, beavers and barns. Precisely because they apply universally to everything, the microscopic laws aren't by themselves going to give us what we need to understand the variegated structure of our world as we encounter it: the world of biology and of social science, and the familiar world of everyday sense. They tell us about the deep, shared substructure of everything, but nothing about the variegated fabric of the world as we encounter it. They tell us nothing about the difference between Sunday and Monday, a rainforest and a river, a rock and a living thing. They tell us little about the dappled fabric of our planet where new regularities arise on temporary scaffolds, creating local pockets of structure. To understand our world over the time scale of evolution in the three and a half billion years since the first appearance of living things, it is not enough to understand the microscopic laws, one has to understand how complex living systems evolve, propagate their own structure, and design their environments. And at the time scale of everyday experience, one has to understand the dynamics of evolved systems of the kind that we find ourselves surrounded by in a world that we encounter. Structures rise and fall, form and disappear, and for the period in which one structure establishes itself, new structures on shorter time scales rise and fall. Such are the rhythms of time: mountains rise and fall against the backdrop of millions of years; empires are built and collapse over centuries; human lives are lived out against the background of mountains and empires; and through each life, seasons come and go.

All of this takes place on a larger canvas: a cosmological background against which the frantic growth and activity on our precious Earth is short-lived and insignificant. And from a physics point of view, the features that are central to our own conception of time are quite special. They depend quite heavily on the thermodynamic gradient, that rising slope from the low entropy past to equilibrium. Without a thermodynamic gradient, there is no distinction between past and future. There are no processes of directed development. A universe in equilibrium is a universe without time as we know it. There is still time in the sense of a dimension in which the universe is extended, but it has none of the distinguishing features of time in our experience.

The warp and weave of the world

In physics, as in philosophy, people often focus on the extremes, but most of the interesting things happen in between. They happen on a scale larger than the very small and smaller than the very large, between the momentary and the permanent, and on the slope between the low entropy past and the high entropy future. The warp and the weave of the world, the patterns with their samenesses and differences, the frozen accidents, the temporary stabilities, the different paces of change, the ebbs and the flows, and the things that rise and fall against different phases of ebb and flow make up the familiar texture of our daily world. They are neither very small nor very big, neither ephemeral nor permanent, neither perfectly regular nor irremediably random. They are mid-size, somewhat permanent, and a potent mixture of regularity and randomness that produces intricate patterning along the temporal dimension. Our own lives are woven into this warp and weave and they are as integral to how things unfold as the growth of trees and the flow of rivers.

It is worth pointing out here that memory and agency—the fact that you remember the past and affect the future—are two sides of the same coin, both rooted in the thermodynamic gradient. Just as you now live among the records of the past, your present actions will leave records, and that means that what you do in the here and now will determine in part the future that you will confront. Your actions disturb the environment in which we live in ways that are both short-lived and enduring. If you make a footprint in sand, it will be gone before the hour. But paint will take longer to fade from a canvas, and the house you built will take decades to crumble. The trees you plant and children you produce will work positively against the natural encroachment of entropy to ensure their own survival long after you are gone. An agent is a system that actively works to arrange and prepare the future it will confront. Squirrels bury nuts for the winter, and beavers build dams that will protect them from predators. Humans look deep into the future and exploit the more durable aspects of their environment to arrange it to their liking.

I want to return in closing to a question that we left out when we were discussing time travel in Chapter 3. One of the reasons that people are fascinated by time travel is that time travel scenarios put extreme pressure on our sense that we can affect the future. In a time travel setting, it doesn't seem like you can kill your grandfather as he lies in his crib even though you stand there, gun cocked, nerves wound tight. It doesn't seem like you can save Abe Lincoln from assassination or keep yourself from the bad decisions that became the bane of your life. In general, there is only one way that things can go that is consistent with the way things stand before your journey. But the idea that we can affect the future (i.e. that there is a range of ways in which events can unfold, any one of which we could bring about) is so basic to our native way of understanding the world that even though there are mathematically and physically consistent ways to describe time travel scenarios, it is hard to make sense of them from the inside.

What that shows is not that time travel can't happen, but that our ordinary concepts of cause and effect, and of agency and action, and more generally, of the differences between past and future don't fare well in a time travel setting, because they presuppose asymmetries that break down along a time travel loop. If we travelled routinely through time, our sense of time would have to be different. That shouldn't be surprising. In the last two chapters we have seen that our own sense of time is very much dependent on the differences between past and future that arise in the context of a thermodynamic gradient. Where those differences aren't present—either because there is no thermodynamic gradient or because one is locally travelling on a time travel loop—our concepts don't apply. In a setting like that, we would need new concepts.

New horizons: beyond space and time

Physics is ongoing, under way, full of surprises, and far from finished. It lurches from crisis to crisis, separated by periods of stability. And the way that it tends to develop through crises is not by adding to an established foundation, but rather by undergoing revolution in its basic concepts. There are reasons for thinking that a new and deeper transformation in our thinking about time is on the horizon. I have spoken throughout this little book as though we live in a universe that obeys the Newtonian mechanical laws. Since the early 1920s it has been known that Newton's laws don't accurately describe the way in which matter behaves below a very small distance known as the Planck Length (about 1.6×10^{-35} m). Below the Planck scale, the laws of quantum mechanics reign. The discovery of quantum mechanics created a rift between our theories of the very small and the very large. The concepts of quantum mechanics do not mesh in a natural way with those of relativity. Attempts to bring them into a single mathematical framework have proved difficult.

The difficulty can be seen by noticing that how things are located in space tells us whether, and by what route, they can affect one another. Something over here can only affect something over there by a signal or sequence of influence that passes through the space between. The same goes for time: something that happened in 1903 can only affect events in 1925 or be affected by events in 1880 by a route that passes through all the times between.

This connection between geometry and causal structure was made explicit with the development of relativity. In relativistic theories, the structure of space and time together encodes the network of causal influence among events. This worked well with Newtonian mechanics and electromagnetism, where events in one volume of spacetime can affect events at another only by influences that pass through the intervening space and do so at finite speed. That is no longer true in quantum mechanics. Events in one part of spacetime are statistically dependent on events in another in a way that cannot be explained by influences passing through the intervening space at finite speed. Different tactics have been proposed to explain the effects, but quantum behaviour is peculiarly resistant to shoehorning into a framework that makes sense relativistically. A number of the most promising attempts to unify quantum mechanics with relativity replace spacetime with fundamental structures that approximate spacetime above the Planck scale. In these theories spacetime is not the fundamental, framing structure of the universe. These matters are completely unsettled. It is one of the frontiers on which philosophically exciting new physics is likely to take place.

References

Chapter 1: Time until Newton

Galileo Galilei, *Dialogue Concerning the Two Chief World Systems* (University of California Press, 1962), p. 186. Translated by S. Drake.

I. Newton, *Philosophiae Naturalis Principia Mathematica*, Bk 1 (University of California Press, 1934). Translated by A. Motte (1729), revised by F. Cajori (1934).

Thucydides, *The Peloponnesian War*, 2.2, 74 (Oxford University Press, 2009). Translated by Martin Hammond and Introduction and Notes by P. J. Rhodes, p. 74.

Chapter 2: From space and time to spacetime: the era of Einstein

A. Einstein, *Annals of Mathematics*, 'On a Stationary System with Spherical Symmetry Consisting of Many Gravitating Masses', *Annals Math.* 40 (1939).

W. Isaacson, *Einstein: His Life and Universe* (Gramercy, 1993).

Chapter 3: Philosophical implications of relativity

F. Arntzenius and T. Maudlin, 'Time Travel and Modern Physics', in E. N. Zalta(ed.), *The Stanford Encyclopedia of Philosophy* (Winter 2013 Edition), <https://plato.stanford.edu/archives/win2013/entries/time-travel-phys/>.

I. Asimov, *Gold: The Final Science Fiction Collection* (Harper Collins, 1995), pp. 276–7.

J. Earman, 'Implications of Causal Propagation Outside the Null-Cone', *Australasian Journal of Philosophy* (1972), 50.
J. Wheeler and R. Feynman, 'Classical Electrodynamics in Terms of Direct Interparticle Action', *Reviews of Modern Physics* 21 (1949), pp. 25–434.

Chapter 5: The time of human experience

D. C. Williams, 'The Myth of Passage', *The Journal of Philosophy* 48/15 (1951), p. 461.

Further reading

Chapter 1: Time until Newton

P. Galison, *Einstein's Clocks, Poincare's Maps: Empires of Time* (Norton, 2004).
P. Kosmin, *Time and its Adversaries in the Seleucid Empire* (Harvard University Press, 2018).
G. W. Leibniz and S. Clarke, *Leibniz and Clarke: Correspondence* (Hackett, 2000).
I. Newton, *Philosophiae Naturalis Principia Mathematica*, Bk 1 (1689); trans. A. Motte (1729), rev. F. Cajori (University of California Press, 1934).
D. Rosenberg, *Cartographies of Time: A History of the Timeline* (Princeton Architectural Press, 2013).
R. Westfall, *Never at Rest: A Biography of Isaac Newton* (Cambridge University Press, 1983).

Chapter 2: From space and time to spacetime: the era of Einstein

A. Einstein, *Ideas and Opinions* (Crown, 1995).
A. Einstein, *Relativity: The Special and the General Theory* (Ancient Wisdom Publications, 2010).
A. Einstein, *Out of my Later Years: The Scientist, Philosopher, and Man Portrayed Through his Own Words* (Philosophical Library/ Open Road, 2011).
A. Einstein, *The World As I See It* (CreateSpace Independent Publishing Platform, 2014).

S. Hawking, *A Brief History of Time* (Bantam Books, 1988).
W. Isaacson, *Einstein: His Life and Universe* (Gramercy, 1993).
P. Wheelwright, *The Presocratics* (Bobbs-Merrill, 1960).

Chapter 3: Philosophical implications of relativity

M. Friedman, *Foundations of Space-Time Theories: Relativistic Physics and Philosophy of Science* (Princeton University Press, 1986).
R. J. Gott, *Time Travel in Einstein's Universe* (Houghton-Mifflin, 2001).
T. Maudlin, *Philosophy of Physics: Space and Time* (Princeton University Press, 2015).
P. Nahin, *Time Machines* (Springer-Verlag, 1999).
P. A. Schilpp, *Albert Einstein, Philosopher-Scientist: The Library of Living Philosophers* Volume VII (Open Court, 1998).
R. Wasserman, *Paradoxes of Time Travel* (Oxford University Press, 2018).

Chapter 4: The arrow of time

D. Albert, *Time and Chance* (Harvard University Press, 2000).
C. Callender, *What Makes Time Special* (Oxford University Press, 2017).
S. Carroll, *From Eternity to Here: The Quest for the Ultimate Theory of Time* (Dutton Adult, 2010).
H. Price, *Time's Arrow and Archimedes' Point* (Oxford University Press, 1996).
H. Reichenbach, *The Direction of Time* (Dover, 1999).

Chapter 5: The time of human experience

D. Buonomano, *Your Brain is a Time Machine: The Neuroscience and Physics of Time* (Norton, 2017).
J. Canales, *The Physicist and the Philosopher: Einstein, Bergson, and the Debate that Changed Our Understanding of Time* (Princeton University Press, 2015).
F. De Waal, *Are we Smart Enough to Know how Smart Animals are?* (W. W. Norton, 2017).
C. Safina, *Beyond Words: What Animals Think and Feel* (Henry Holt and Co., 2015).

Chapter 6: The big picture and new horizons

S. Carroll, *Something Deeply Hidden: Quantum Worlds and the Emergence of Spacetime* (Dutton, 2019).

J. Hartle, *The Quantum Universe: Essays on Quantum Mechanics, Quantum Cosmology, and Physics in General* (World Scientific Publishing Company, 2020).

N. Huggett, K. Matsubara, and C. Wuthrich(eds), *Beyond Spacetime: The Foundations of Quantum Gravity* (Cambridge University Press, 2020).

C. Rovelli, *Reality is Not What it Seems: The Journey to Quantum Gravity* (Riverhead Books, 2017).

L. Smolin, *Three Roads to Quantum Gravity* (Basic Books, 2017).

L. Smolin,*Time Reborn: From the Crisis in Physics to the Future of the Universe* (Houghton Mifflin Harcourt, 2013).

General

F. Arntzenius and T. Maudlin, 'Time Travel and Modern Physics', in E. N. Zalta(ed.), *The Stanford Encyclopedia of Philosophy* (Winter 2013 Edition), <https://plato.stanford.edu/archives/win2013/entries/time-travel-phys/>.

C. Callender, 'Thermodynamic Asymmetry in Time', in E. N. Zalta(ed.), *The Stanford Encyclopedia of Philosophy* (Winter 2016 Edition), <https://plato.stanford.edu/archives/win2016/entries/time-thermo/>.

B. Dainton, *Time and Space* (Routledge, 2021).

J. Earman, C. Wüthrich, and J. B. Manchak,'Time Machines', in E. N. Zalta (ed.), *The Stanford Encyclopedia of Philosophy* (Summer 2020 Edition), <https://plato.stanford.edu/archives/sum2020/entries/time-machine/>.

B. Greene, *The Fabric of the Cosmos: Space, Time, and the Texture of Reality* (Knopf, 2004).

R. Le Poidevin,'The Experience and Perception of Time', in E. N. Zalta (ed.), *The Stanford Encyclopedia of Philosophy* (Summer 2019 Edition), <https://plato.stanford.edu/archives/sum2019/entries/time-experience/>.

N. Markosian, 'Time', in E. N. Zalta (ed.), *The Stanford Encyclopedia of Philosophy* (Fall 2016 Edition), <https://plato.stanford.edu/archives/fall2016/entries/time/>.

C. Rovelli, *The Order of Time* (Riverhead Books, 2018).

A lot of the philosophical discussion of time, including the paradoxes of time travel and contested questions about the philosophical implications of physics, is carried out in journal articles. *The Stanford Encyclopedia of Philosophy* has a number of entries relevant to the topics covered here. These entries are excellent surveys and come with detailed bibliographies that are updated regularly.

S. Savitt, 'Being and Becoming in Modern Physics', in E. N. Zalta(ed.), *The Stanford Encyclopedia of Philosophy* (Fall 2017 Edition), <https://plato.stanford.edu/archives/fall2017/entries/spacetime-bebecome/>.

N. J. J. Smith, 'Time Travel', in E. N. Zalta (ed.),*The Stanford Encyclopedia of Philosophy* (Summer 2019 Edition), <https://plato.stanford.edu/archives/sum2019/entries/time-travel/>.

Index

For the benefit of digital users, indexed terms that span two pages (e.g., 52–53) may, on occasion, appear on only one of those pages.

A

B

C

D

E

Q

R

S

T

U

V

W

THE HISTORY OF LIFE

A Very Short Introduction

Michael J. Benton

There are few stories more remarkable than the evolution of life on earth. This *Very Short Introduction* presents a succinct guide to the key episodes in that story - from the very origins of life four million years ago to the extraordinary diversity of species around the globe today. Beginning with an explanation of the controversies surrounding the birth of life itself, each following chapter tells of a major breakthrough that made new forms of life possible: including sex and multicellularity, hard skeletons, and the move to land. Along the way, we witness the greatest mass extinction, the first forests, the rise of modern ecosystems, and, most recently, conscious humans.

www.oup.com/vsi

RELATIVITY

A Very Short Introduction

Russell Stannard

100 years ago, Einstein's theory of relativity shattered the world of physics. Our comforting Newtonian ideas of space and time were replaced by bizarre and counterintuitive conclusions: if you move at high speed, time slows down, space squashes up and you get heavier; travel fast enough and you could weigh as much as a jumbo jet, be squashed thinner than a CD without feeling a thing - and live for ever. And that was just the Special Theory. With the General Theory came even stranger ideas of curved space-time, and changed our understanding of gravity and the cosmos. This authoritative and entertaining *Very Short Introduction* makes the theory of relativity accessible and understandable. Using very little mathematics, Russell Stannard explains the important concepts of relativity, from E=mc2 to black holes, and explores the theory's impact on science and on our understanding of the universe.

www.oup.com/vsi

NOTHING

A Very Short Introduction

Frank Close

What is 'nothing'? What remains when you take all the matter away? Can empty space - a void - exist? This *Very Short Introduction* explores the science and history of the elusive void: from Aristotle's theories to black holes and quantum particles, and why the latest discoveries about the vacuum tell us extraordinary things about the cosmos. Frank Close tells the story of how scientists have explored the elusive void, and the rich discoveries that they have made there. He takes the reader on a lively and accessible history through ancient ideas and cultural superstitions to the frontiers of current research.

> 'An accessible and entertaining read for layperson and scientist alike.'
>
> Physics World